REVIEWS OF SPEECHLESS

SUSAN JAMES

In a world where effective communication has never been more critical, the world's employers report this as one of the major skill gaps. Students in the UK are not adequately equipped for the demands of their future lives and yet more young people than ever are experiencing mental health issues generated by feeling overwhelmed by the demands that they experience at school age. Even more worrying, the link the author forges between a lack of ability to self-talk and likelihood of crime-related activity in the future provides a serious pause for thought.

Rosemary Sage's *Speechless* gives voice to the current concerns our society faces with regard to education and how best to prepare our young people for the future. It acknowledges that society, and therefore language, is in a constant state of change. At the heart of this text are two key principles: educators should be given time to understand the circumstances of each child; students perform best when they are given opportunities to learn in a way that is participatory. Understanding the value and importance of language, teaching pupils to function as part of a community, will ensure they can understand and take ownership for their thoughts and actions. This offers society a positive way forward.

This book is an energised, wide-ranging, intelligent, humorous human journey. It sweeps over time, history and a vast array of cultures, societies, studies and ideas in order to provide the reader with a rich but readable response to the challenges that educators, parents and young people face today and in the future.

Susan James is a teacher of English Language and Literature, as well as a Pastoral Deputy Head at Cheadle Hume School. She has been a Head of Lower School at the Manchester Grammar School; a Drama and PSHE teacher; a Form Tutor and an Academic Mentor for English

PGCE Students and NQTs. She achieved a Distinction for her MEd in Educational Leadership at the University of Buckingham, UK. Following her work on developing resilience in school pupils, Susan has worked to create a new curriculum for Lower School boys at Manchester Grammar School, from 2018. She is a School Governor, with many community interests and is particularly focused on the holistic development of children for their futures in the new Industrial Revolution (4).

SHERRY BRANDENBURG BENT

The speed of technological advances has ricocheted the educational establishment into seeking solutions and finding answers for the drastic impact it has had on our ability to teach students. Too many hours spent on iPads, phones and on social media has resulted in a lack of focus, underdeveloped oral communication skills and a detachment in responding appropriately to teachers in the classroom. Technology has created a void in how effective we are at reaching our students' potential intellectual growth and development.

This book is a wake-up call for all teaching practitioners by exposing the damaging effects of too much time spent absorbed in virtual platforms and the implications this has on equipping students for tomorrow's world. Most importantly, this book highlights the vital necessity we face in schools to improve classroom dialogue and verbal communication so students can effectively navigate a more complex and multi-faceted job market.

The future of our education system's success lies in developing strong communication abilities that transcend both technology and a rapidly changing global template. Developing interpersonal skills and the ability to respond both orally and in written form is paramount to success in all fields of work. The role of schools to hone oral language, as well as developing a high standard of written language, is vital to 21st-century life.

Language and vocabulary are developed in early childhood and are built throughout years of schooling and throughout adult life. Children are now entering school with insufficient language and vocabulary development which affects their progression in both reading, writing and all areas of the curriculum. Thus, the problems become compounded throughout the primary school years and beyond.

Verbal communication, discussion and dialogue must be at the forefront of education. This book should be required reading in every teacher training college and by all educational professionals.

Speaking, listening, and responding intelligently must be at the heart of the information age.

Sherry Brandenburg Bent *(MA Education in Curriculum Development and Supervision) has taught in primary and middle schools in the US for twenty years before coming to England on a Fulbright Scholarship in 1991. Since living in England, she has taught all primary grade levels, as well as at the foundation stage and in secondary schools. She was a literacy consultant for ARK Schools for five years before starting Brandenburg Bent Tutorials Ltd in 2014. Her school specialises in supporting primary students, by building strong literacy and mathematics skills, as well as focusing on problem solving and communication competencies, especially for students sitting for 11+ exams. Sherry is also chairman of a London-based charity that specialises in teaching reading and building literacy skills in disadvantaged children.*

SPEECHLESS

Issues for Education

Rosemary Sage

University of Buckingham Press

The University of Buckingham Press,
51 Gower Street, London WC1E 6HJ
info@unibuckinghampress.com I www.unibuckinghampress.com

Contents © Rosemary Sage 2020
The right of the above author to be identified as the author of this work has been asserted in accordance with the Copyright, Designs and Patents Act 1988. British Library Cataloguing in Publication Data available.

Print ISBN 9781789559330
Ebook ISBN 9781789559323
Set in Times.

Speechless

Issues for Education

Rosemary Sage, University of Buckingham Press

DEDICATION

This book is dedicated to today's courageous children who are growing up in a complex, competitive, confusing world, where they need effective communication and confidence to survive. Have you got a question about this? Reach for your smartphone. Google it. We are deluged with easy-access information at our fingertips 24/7, with the result that our brains are getting weaker, leaving us with the attention span of a goldfish, according to researchers! Those growing up with abundant technology – now referred to as the '*Goldfish Generation*' – could be at a greater risk of memory decline. Academics from Oxford, London, Harvard and Sydney have all found that our brains are changing because of the Internet. Significant reductions in grey matter in the orbitofrontal cortex of the brain, implicated in impulse control and decision-making, have been discovered in research studies. Smartphones are replacing our ability to remember facts and to communicate orally face-to-face.

The other problem with being constantly connected is that we have become easily distracted. Research published in the *Journal of Communication* found we spend an average of 19 seconds focusing on something online before we flick to the next thing, with 75% of all on-screen content being viewed for less than one minute. A distracted mind will not take in information deeply enough to engage with what we are processing. For example, hyperlinks – even when not clicked on – reduce information-processing capacity for the present content as the mind is on where to go next. Research at the Technical University of Denmark suggests the global attention span is narrowing due to the huge amount of material that we are all faced with daily – labelled as '*information overload*' or '*info-besity*'.

This research has implications for teachers or trainers tasked with the job of helping people to learn. The book is an update of others on the subject of the communicative process for learning, written over a decade ago by the author. It reflects the research of a number of European

Commission projects on the subject of *education, communication, learning and the workplace*. In a plural society, where many languages are spoken, reflecting the culture, values and thinking of a great range of people from across the world, the importance of understanding how to communicate to establish effective relationships and develop higher-level thinking for learning is crucial. *Speechless* aims to build awareness of the issues and is of interest to anyone wishing to develop their own survival abilities and for teachers who need for up-to-date understanding so that students can learn successfully.

PREFACE

Professor Barnaby Lenon

In recent years teachers have become reluctant to use the phrase 'educational failure' for fear that it will discourage children. Long gone are the days when end-of-term reports castigated mainly boys for being 'dim', for slovenly work and appalling manners. Think of Winston Churchill at Harrow, whose exasperated housemaster wrote to his parents about him being '*so regular in his irregularity that I really don't know what to do*'. The emphasis now is on praise and encouragement.

Part of me thinks that this is fair enough. Those ancient reports were too often only written for the entertainment of other teachers and were often cynical in tone ('*the improvement in Fabian's handwriting has revealed his inability to spell*' or '*for this pupil, all ages are dark*'). They rarely considered their impact on the children and often they were plain wrong, like Albert Einstein's teacher ('*He will never amount to anything*') or that of Gary Lineker ('*He must devote less of his time to sport if he wants to be a success. You can't make a living out of football*')!

The notion of failure has been happily leavened by the concept of progress. If you start at a low base but make good progress over time – that is success, even if your mark in the final exam is low. Now most school reports have a progress and attainment grade. That is sensible.

The trouble with the modern, kindly approach is that it can obscure real educational weaknesses (let us not call them failures). In England, in 2019, 35% of pupils taking English and maths GCSEs (for many the most important school exam they will ever take, the culmination of 11 years of study) failed to get a bare pass (grade 4). The OECD PISA tests place England lower in the international rankings than we would wish, and Wales and Scotland are worse. The 2015 PISA tests had the UK 13th out of 72 countries in science, 17th in reading and 26th in maths. British 15-year-olds' maths abilities are more than two academic years behind 15-year-olds in Singapore and China. The OECD equate 30 PISA test points to a year of additional schooling; so if we take PISA 2015 maths for example, England scored 493, Singapore 564 – 2.4 years' difference.

In maths, our lowest-achieving pupils are much worse than the lowest-achieving pupils in many other countries. In PISA 2015, 22% of UK pupils did not reach level 2 in maths, which means that they *'cannot solve problems routinely faced by adults in their daily lives'*. Although the best pupils do well, our tail of low-achievers is longer than in other countries. The priority is therefore to improve the basic standard of schooling for less motivated or less able students.

In fact, 18–24-year-olds in England have weaker literacy and numeracy than 55–65-year-olds – almost the only country in the world where this is true. The 50 per cent of the population who do not go to university find themselves in a system of vocational education which is notoriously weak by the standards of most other European countries.

The concepts of success and failure are made more complicated by the fact that no one can be completely confident about the aims of education. In the past few years, the Minister for Schools in England has vigorously promoted mathematics and levels of *'factual'* knowledge. I do not disagree with him, but those who are keener than him on arts, creative subjects and 20th-century competencies will take a very different view of success and failure. One thinks of the CBI's regular condemnation of school-leavers who, notwithstanding their grades, cannot speak well or operate successfully in teams. One feels challenged when reading the chapter in this book about the strengths of schools in Cuba, one of the poorest countries in the western world.

Notwithstanding the endless rhetoric in the UK about social mobility, half the population will always have below-average cognitive ability and school exam results. But if we are to overcome the divisions in society revealed by recent political events, all need to feel valued. That is why we must get back to a more balanced appreciation of those who perform essential jobs, albeit jobs of the hand or heart rather than intellect. We should stop talking about social mobility as simply a way of *'rescuing'* people from working-class backgrounds and place more emphasis on valuing the full range of worthwhile occupations.

Arguments about social mobility are normally based on exam results or incomes, not the value to society of different occupations. School accountability measures give little weight to skills other than exam performance. As this book suggests, we need to find a place where those of below-average cognitive ability can nevertheless earn a good living and feel satisfied with their lives. Only then can we build a more productive, happier and less divided society.

INTRODUCTION

Riccarda Matteucci

In the world of education, teachers, trainers, administrators, policy officials and well-meaning adults often gather together with an intention to implement and improve student performance. They discuss instructional techniques based on research studies, first-hand experience and any other data gathered. Best practices are evaluated for the greatest impact that innovatory methods have on students to enhance performances. Sometimes traditional activities are dusted off and named differently to create novelty and expectation. However, what is important to understand is that education goals will change in line with world developments. A traditional focus on giving students knowledge and skills to become compliant employees does not fit a situation where routine jobs are rapidly disappearing due to intelligent machines, so requiring more creative human competencies for higher-level work. We might ask:

'How can we reach all students? How can we prepare them for the 21st-century workplace? How can we support them to be lifelong learners? How can we win over systematic discrimination in plural societies?'

All answers can be reduced to one word: Communication.

Professor Rosemary Sage has indicated in her research over many years that time can be wasted seeking the latest teaching gimmick when the priority is to communicate and connect with students as the key to effective learning (Sage, 2000–2020). Communication means all aspects involved in transmitting a message, which are not only verbal, but also non-verbal, such as tone of voice, manner and gestures, as well as connotation (*feelings/ideas suggested by words rather than the meaning*), along with the context in which the event occurs.

For example, the word *'discipline'* (self-control) has the unhappy connotation of punishment and repression.

Neuroscientists have clarified the role of *collaboration in communication*, with research showing that when we connect with other ideas there are multiple benefits for our brain and life progress. When people collaborate, the medial orbitofrontal cortex and the frontoparietal network are activated to support executive functions. These areas are known as the *'social brain'*. When we work together our brains are charged with the complex task of making sense of each other's thinking, and so learning to interact is crucial. Social cognition is an important area of current neuroscientific investigation.

Educators need to convince students that school and later college success requires working with their peers to create a community based on shared intellectual interests and common aims. We need to impart to students how they can achieve working together and, in some cases, teach teachers *what* and *how* to do it. Teachers are not taught about learning through the process of communication, with narrative language essential for developing higher thinking, creativity and problem-solving.

TALK AND SHARE

Educators know in theory the importance of sharing ideas, but many still see no role for collaborative learning in schools, where the focus is on individual success in standard national tests. Students give up learning when they find it difficult and think they are alone in their struggle. A change takes place when, working together, they discover that everybody finds some or all of work a major challenge. If there is no real communication between people in an educational setting, how can we encourage the teaching-learning process? Transmitting knowledge through cooperative collaboration is the key to development. Rosemary Sage specifies that appreciating this represents a critical moment in student awareness. Learners then know what is involved in developing knowledge, skills and attitudes, and that obstacles in their acquisition are commonplace and can be negotiated. Her view is that attitudes change when students have the chance to connect ideas. Linking with another's thoughts develops a higher level of understanding. Whatever the subject matter, when students cooperate they integrate their knowledge and experience as the basis of creative thinking.

Interesting findings come from results of a large-scale testing programme, such as the Programme for International Student Assessments (PISA, 2016). There are many possible social and emotional reasons for this. While PISA reveals large gender differences in reading, in favour of 15-year-old girls, the gap is narrower when digital reading skills are tested. Indeed, the Survey of Adult Skills suggests that there are no significant gender differences in digital literacy proficiency among 16-to-29-year-olds. Boys are more likely to underachieve when they attend schools with a large proportion of socio-economically disadvantaged students. Girls, even high-achieving ones, tend to underachieve compared to boys when they are required to think like scientists. This is when they are asked to formulate situations mathematically or interpret phenomena scientifically. Parents are more likely to expect their sons, rather than daughters, to work in a science, technology, engineering or mathematics field – even when their 15-year-old boys and girls perform at the same mathematics level. These international tests, carried out worldwide, show that boys achieve better than girls in mathematics in 38 countries out of 57 in the PISA scheme.

Our obsession with testing and comparing students, schools and nations is leading to anxiety. Rosemary Sage points out the importance of expressing our feelings through talk in order to balance our thinking, opinions and views. Our favoured way of communicating is via electronic devices, but it is face-to-face exchanges that bond people together and provide the support they need when times are difficult and they lose confidence. The PISA team (2012, 2016) presented results in mathematics, with the gap in achievement between sexes initially explained by the lower confidence of girls due to the '*gender*' factor. In reality the difference in the lower performance of girls was due to '*confidence*'. Girls became more anxious when they took the individual mathematics test. This phenomenon is now well established and recognised. It should make educators reflect before basing decisions on test performance, as Rosemary Sage has been underlining for a long time.

Going back to PISA test studies, the team conducted another assessment to reduce inequalities, adding to the usual individual mathematics test a collaborative problem-solving one. The students did not collaborate with each other but with a computer agent. They had to take on the agent's ideas, connect with these and build on them to collaboratively solve complex problems. Instead of producing individual results, students were required to produce combined responses. In the test of collaborative problem-solving, administered in 51 countries, girls outperformed boys in every one. Two more noticeable results occurred. There were no significant differences in outcomes between advantaged students and in some countries diversity boosted performance. The team also found that in some countries indigenous students reached higher levels when in schools with large numbers of '*immigrants*'. This result suggests that diverse learner communities help students become better collaborators. PISA tests reveal the potential of collaboration not only for girls or students from different countries but for all thinkers and learners. When we connect with someone else's ideas, understanding and perspective are enhanced.

Collaboration is vital for learning, for college success, brain development and creating equitable outcomes. Beyond this, it is beneficial to establish an interpersonal *verbal* connection, especially in times of conflict and need. Rosemary Sage's ideas on education show her genuine enthusiasm for all to succeed. What does it take to make education prosper? Schools are constantly trying new things to improve student outcomes that sometimes work but too often fail. Educators have been testing and refining what it takes to achieve sustainable change and transform schools, in recent years, as world competition in everything becomes fierce. For innovation to blossom five conditions are needed: ***conviction***, ***clarity***, ***capacity***, ***coalition***, and ***culture*** (the Hechinger Report, Butrymowicz & Kolodner, 2019). These are referred to as a holistic approach, and lay the groundwork for successful school transformation (Sage 2017).

Conviction is the need to change with a specific transformation plan to achieve it;

Clarity is keeping in mind the goals of the change and strategies to achieve them;

Capacity refers to finances, staff expertise and time to dedicate to the initiative;

Coalition is support by leaders, educators, families and community members;

Culture is learner-centred: collaboration and risk-taking for continuous improvement.

The mission is to support communities to build and spread equitable learning environments. For personalised, competency-based education, Rosemary Sage identifies three essential elements for successful new schools:

Strong vision from input of school leaders, teachers, students, parents and community;

Collaborative culture that empowers individuals and encourages risk-taking;

Transparency about what is expected from students in the 'voice' needed to convey hopes and requirements for learning. Also, what roles should adults play in this vision?

The core idea in *Speechless* is that schools relying on educators to do all the teaching is not viable any more. Far-reaching school changes happen when the wider community is involved, with most learners motivated by the relationships within this. Rather than top-down mandates, widespread support for initiatives depends on backing up an innovative culture through collaborative action. The normal ups and downs of life must not discourage educators to discontinue innovation. They need to establish the right foundation from the start, which depends on the *communication*, *collaboration* and *cooperation* of all stakeholders.

Schools also need to incorporate health and well-being activities into daily routines. Peter Gray, Psychology Professor at Boston College, says that the 1950s were the high point in the United States, and I would add in other countries, too, such as England and Italy. Family members could keep an eye on their children playing outside, in the courtyard or street, by watching through a window. Children played safely and happily with friends, learning how to navigate the world by interacting with one another.

TALK WITH PASSION

The activities of 1950s children were unstructured and unsupervised by adults – the latter were only concerned that no one was hurt. The youngsters generated their activities and rules of conduct. If someone broke these, the children themselves determined the consequences. Through such experiences they developed resilience, self-determination and problem-solving. Research showed that the competencies needed for these activities were gained by being unsupervised, developing positive mental health, social and emotional skills, intrinsic motivation and creativity (Gray, 2017). This could be considered a relic of the past. Modern factors, such as fractured family dynamics, economic and academic anxiety, fear of strangers and vehicular traffic, have all put unsupervised '*play*' at risk. If we compare the exposure of daily risks of social media, would you consider youngsters are safer today?

The periods that children spend freely have reduced enormously. Gray states that '*childhood is turned from a time of freedom to a time of résumé-building*' (2017, p. 9). He connects a decline of free play to the rise in mental-health issues. Paediatricians advocate protection of a child's unstructured time because of its numerous benefits, including the development of speaking and motor abilities, which may have lifelong aids for learning and the prevention of obesity, hypertension and type 2 diabetes. They see an out-of-balance lifestyle for today's youngsters, and Gray supports this by emphasising the learning value of unstructured play. We must convince children to abandon the mobile phone in favour of free time with their friends. It does not matter if they end up with some bruises. They can relate and explain how these happened to their carers, giving opportunity to practise narrative thinking and language structure.

We have developed a technology-dominated world, which has encouraged child surveillance by home-cams, GPS trackers and web-watchers at home and, increasingly, in schools across the world. Although, used wisely, surveillance technology can be a blessing, too often the easy power and convenience it brings spark worry over petty matters. Even worse, it pushes people away, by allowing us to covertly circumvent the basic elements of a healthy relationship. As Rosemary Sage says: '*conversation, discussion and trying to understand different people's viewpoints is the progressive and more successful way of doing things*'. Emotionally connecting and communicating with children is more effective than coercion and control. Those who monitor others usually insist they want safety and peace of mind, but this controlling action is damaging relationships. Gaining respect and trust is harder work than relying on technology, but when mutual it increases the chance of cooperation. One must have conversation and negotiation in any relationships that matter. Communicating effectively teaches and trains students to self-regulate and take control and initiative over all their activities.

TALK, COMMUNICATE, RELATE AND RESOLVE MUST BE THE MANTRA OF TODAY

The Global School Play Day started in 2015. It enabled an entire day to be devoted to play, as teachers recognised the importance for children to communicate and connect with one another in creative activities. It should be adopted worldwide. Very wisely, Rosemary Sage suggests that there is no change without **communication** and **collaboration**. Therefore, the five Cs have been joined by two more. The soil is fertile to nurture whatever initiative is planned to make a noticeable change and ensure that it is sustained and maintained. Let us start at once and enjoy the challenge!

REFERENCES

Butrymowicz, S. & Kolodner, M. (2019) *The Hechinger Report on Innovation and Inequality. hechingerreport.org/special-reports/higher-education/* The Hechinger Report explores the innovations at the core of today's cutting-edge schools.

Gray, P. (2017) *Free to Learn*. New York: Basic Books PISA (Programme for International Student Assessment) www.oecd.org/pisa/pisa-2015-results-in-focus.pdf

Sage, R. (2000) *Class Talk* Stafford: Network Educational Press.

(2000) *The Communication Opportunity Group Scheme*. Leicester: University of Leicester

(2003) *Lend us Your Ears: Understanding in the Classroom*. Stafford: Network Educational Press

(2003) *Support for Learning*. Exeter: Learning Matters

(2004) *A World of Difference: Developing School Inclusion*. Stafford: Network Educational Press

(2004) *Silent Children: Approaches to Selective Mutism*. Leicester: University of Leicester (French & Japanese)

(2004) *Support for Learning*, 2nd Edition. Exeter: Learning Matters

(2004)*Tackling Inclusion in Schools*. Victoria, Australia: Hawker Brownlow Education

(2006) *Communication, Emotion and Learning: Handbook for SEBD*. London: Continuum Press. Editor

(2006) *Communication and Learning*. London: Sage

(2006) *Assessment and Teaching the Communication Opportunity Group Scheme*. Leicester: University of Leicester

(2007) *Inclusion in Schools: Making a Difference*. Stafford: Network Educational Press

(2007) *COGS in the Classroom*. Leicester: University of Leicester

(2010) *Meeting the Needs of Students with Diverse Backgrounds*. London: Network/Continuum

(2011) *Key Competencies in the Context of Contemporary European Education (Editor)*. EU: Commenius

(2012) *Issues regarding Adult Literacy and Employment (Editor)* EU: Commenius

(2014) *Policy for Educators' E-portfolios to Improve Professionalism (Editor)* EU: Commenius

(2017) *Paradoxes in Education (Editor & contributor)*. Rotterdam: SENSE

(2019) *The Robots Are Here: Learning to Live with Them. (Editor & contributor)*. London: University of Buckingham Press

PROLOGUE:
A NEW CONVERSATION
ABOUT EDUCATION

INTRODUCTION

Children find face-to-face conversation '*too much effort*', preferring to watch YouTube, an Ofcom report (2019) claims. The media watchdog says that many 4-to-16-year-olds would rather watch screens alone than meet with friends or pursue sports and hobbies. These are more exertion than they feel like expending. Are you speechless? We should all be shocked, reflecting on the consequences for the mental, physical and emotional health of these youngsters. Studies show that many learners lack the conversational moves necessary to shift home *dialogue* into class *monologue*, with educators transmitting knowledge in large chunks of talk or text (Sage, 2000). In 2002, the Basic Skills Agency (BSA) reported 50% of school children with significant communication problems. This situation appears to have worsened two decades later, according to numerous sources.

Blogs on *Huffington Post* say that youngsters increasingly enter school with limited experience of face-to-face talk, which hinders learning and relationships. Mehrabian (1971) suggests 93% of affective message comprehension depends on non-verbal aspects. Therefore, it is not surprising that students find extended speaking difficult in a world where remote '*tech*' communication is more common than face-to-face '*talk*'. The BSA definition of communication difficulties (report above) was limited to word sounds and sentence structures rather than how these are connected for successful message transmission. This requires *information* and *opinion gaps* to be bridged, using *reference*, *inference* and *coherence* tactics to achieve comprehension. The report emphasis was on language *form* rather than its *content* and *use* for various purposes. Communication is rarely taught formally, as you might find in the 10 top nations in educational league tables. Is this a reason for the Flynn studies (2017) showing a decline in general intelligence quotients (IQ) because less speaking is used today for promoting higher-level thinking? Several studies also suggest that IQ has been declining this century due to endocrine disruption, from chemical pollution, affecting the thyroid hormone with devastating effects on cognitive development (Demeinex, 2017).

An analysis of some 730,000 IQ test results, by researchers from the Norwegian Ragnar Frisch Centre for Economic Research, reveals the Flynn effect hit its peak for people born in the mid-1970s and has significantly worsened ever since. This is the most convincing evidence yet of a reversal of the Flynn effect, but of course it is only one major study. If you assume their model is correct, the results are impressive and provide information for reflection. The research team sourced data from IQ test scores of 18-to-19-year-old Norwegian men, who took the tests as part of compulsory military service. Between 1970 and 2009, three decades of men (*born between 1962 and 1991*) were conscripted, giving over 730,000 IQ test results. What these show is that a turning point for the Flynn effect occurred for the post-1975 birth cohorts, equivalent to 7 fewer IQ score points per generation. The Flynn research on IQs of British teenagers, nearly a decade ago, observed a similar fall in test scores.

Lifestyle could well be behind these lower IQs, perhaps due to the way children are educated, how they are brought up and the things they spend time doing (*types of play; book reading; screen activity, etc.*). Another possibility is that IQ tests have not adapted to accurately quantify an estimate of modern intelligence – favouring forms of taught reasoning that may be less emphasised in present education and lifestyles. Researchers distinguish between *fluid intelligence* (*able to see new patterns and use logic to solve novel problems*) and the *crystallised* type (*things taught and trained*) (Cattell, 1963). Small studies, shown below, suggest problems in *fluid* thinking exist, which is a vital competence for solving complex challenges.

A Medical Research Council project investigated 300 children, who were average on intelligence tests but failing in school (Sage, 2000). Their thinking and narrative language structures were found to be deficient in each case. These youngsters could chat with friends happily in the playground, but in class showed an inability to assemble extended talk to make meaning, or to present ideas to others in a coherent structure. This aspect is not assessed in standard intelligence tests. Frequently, employers state that nowadays many workers find it difficult to string words together for processing and producing ideas.

They fail to process or give instructions effectively and, therefore, are powerless to give clear information in exchanges. Of course, excellent communicators and thinkers exist. However, ineffective interaction is a rising problem, especially in a world requiring us to be smarter for collaborating effectively with other experts in solving urgent problems. People can be fluent in *dialogue* speech, where they have some control, but when faced with *monologue,* with *one* person in charge, there is less opportunity to influence direction. Therefore, they frequently have trouble in processing and assembling ideas for making meaning. These concerns are generally not acknowledged or evaluated in formal or informal learner assessments, and rarely receive sufficient attention in teacher training.

Language is a defining factor for students at any level. Those lacking *clear, connected speech sequences* do not have the narrative thinking and linguistic forms to acquire the secondary language activities of written literacy and numeracy easily, or express feelings for gaining emotional support. Furthermore, this absence constrains the development of higher-level thinking and reflective action, causing behaviour problems in schools. Students would rather be thought '*bad*' than '*dim*', research suggests (Sage, 2000). Limited understanding of cause and effect underlies the escalating numbers of serious crimes of 5-to-15 year-olds (*Magistrates in the Community Project – MIC**). The individual danger is restricted personal and academic performances. In respect to society, limited social communication and cooperation makes the recourse to crime more likely. We ignore the fostering of the good side of *human nature* at our peril, and teachers are vital in supporting positive student actions.

WHAT DO STUDENTS NEED TO LEARN IN THIS NEW TECHNOLOGY AGE?

What should we encourage in today's learners for their futures alongside intelligent machines, now increasingly taking over daily routines? There has been a push across the world to offer every student hands-on

---

*The Magistrates in the Community (MIC) project is an initiative that has developed over the past 20 years to increase public awareness of the role of magistrates in the criminal and civil justice system. Through various activities, magistrates help people understand the consequence of their actions. Magistrates attend primary and secondary schools, sixth- form colleges, community groups and employers. They give a presentation and involve participants in various scenarios which help them to think effectively in challenging situations. (www.magistrates-association.org.uk/About-Magistrates/Magistrates-In-The-Community)

computer science and mathematics classes to make them job-ready from day one. Amazon aims to teach 10 million children a year through its Amazon Future Engineer programme. Facebook, Microsoft, Google and others have similar projects. Computer-programming languages like Scratch and initiatives such as Hour of Code are popular with youngsters. Programming can be fun for those that enjoy it, but what about others who are less enthusiastic? It is unlikely that we will be programming in the way we do now in years to come. Machine learning, or artificial intelligence (AI), is very different from giving a computer detailed, step-by-step instructions in coded programmes. Instead, machine-learning algorithms are fed and the large data sets and programmes themselves construct the models that do the analytic work infinitely quicker than human beings.

For example, Google Translate used to involve 500,000 lines of code but now employs only 500 in a machine-learning language. The student challenge is not only knowing programming language but understanding how computer-constructed models work mathematically, to be tested, reviewed and refined for real purposes.

What matters then for people in the future? The technical side of work is mathematical – linear algebra, calculus, probability and statistics. Mathematics, therefore, will be a significant daily requirement. Algorithmic thinking, however, does not just come from computer coding, and appropriate learning experiences are via cooking, sewing, knitting, art, music, drama and sport, etc. – all of which involve algorithms. Computer programming, alone, encourages closed-world building, with some people loving the magic of putting something together instruction-by-instruction and then playing in the virtual space they have built. However, this is far from today's tech world. Programmes now create tools that interact with challenging realities.

PROGRESS HAS SLOWED BECAUSE PEOPLE SKILLS ARE LACKING

It is thought that our attraction to building insular, virtual worlds has hindered understanding of how tools can function in real-life challenges. In this century, for better or worse, progress is not what it used to be. By 1970, all key technologies of modern life were in place: *electricity* to power home systems for cooking, heating and cleaning; *sanitation* for improved health; *mechanised machines* for agriculture, like mowers and tractors; *highways* for speedy transport; *air travel* to all countries; and *telecommunications*, with the Internet connecting people remotely and allowing them to seek information. However, since the 1970s the pace of the last century has considerably slowed. The only notable development is the increase in computing power and the sophistication of mobile devices. In most other ways the lives of people are similar to 50 years ago. This is positive from the viewpoint of giving people time to adapt and consolidate, but forward movement has been less than predicted.

Consider consumer robotics. There is huge potential for robots to help with housework, teaching, medical and social care, as well as leisure and entertainment, but companies working towards these goals have been rapidly closing. In March 2019, Jibo, the social robot maker, after raising $73 million in venture capital, ceased trading. In April 2019, Anki, the toy robot maker, shut down, despite obtaining $182 million for development. The only successful venture has been iRobot's Roomba, a vacuum cleaner operating since 2002. With regard to space investigation, in 2007 the XPRIZE Foundation offered $30 million, funded by Google, to commercial teams in order to land a robotic rover on the moon. However, no team has been able to meet the deadline. They all had difficulties in raising finance for launch contracts and finding the relevant human expertise to develop this invention. Seattle-based Spaceflight Industries are pioneering low-cost ride-sharing into space for small satellites, but the cost for larger ones and probes is sky-high and beyond their means.

One technology achievement is smartphones, and we have had them long enough to review their dangers, such as fake information, cyber-bullying and personal information hacking. Deepfake technology manipulates video and audio for malicious purposes, whether for promoting violence or defaming personal reputations. The Internet Watch Foundation revealed in September 2019 that children as young as 11 years are being persuaded to film and broadcast sexually explicit videos of themselves, in a worrying trend that sees over 100 cases a day. Also, the National Society for the Prevention of Cruelty to Children (NSPCC) published figures saying the number of grooming cases in which paedophiles had contacted children online had risen by a third in 2019. The Climate Psychology Alliance (CPA) has announced rising numbers of students treated for 'eco-anxiety' – fear of the future. Apocalyptic warnings by activists, such as Extinction Rebellion and Resisting Whiteness, have prompted more young people to seek help because they feel they cannot cope with life pressures. Dr Claire Weekes (pioneering psychoanalyst) developed the *anxiety code,* focusing on six words designed to calm the mind: *face, accept, float, let time pass* (Hoare, 2019). We need to separate fact from fiction, as these forecasts of doom shape what we believe about the world and its people and lead to depression and caution.

Self-driving and flying cars, augmented-reality glasses, gene therapy and nuclear fusion are all still in prototype and have been for a while. Accountability, when things go wrong, is a big consideration today, and the greater complexity of present innovations require human skill sets not yet properly developed. There are hard problems to crack, like zero- and negative-emissions technology, essential for public health. Not only is finance needed for this, but communicative and creative human exchanges are also required to make these developments possible.

THE WORLD HAS IMPROVED, BUT TECHNOLOGY REQUIRES AWARENESS

We have been able to solve huge problems during the 20th century. Economic data shows how the world has improved. In 1919, for example, 67.1% of the world's population was living in extreme poverty, whereas one hundred years later this proportion has dropped to just 10.6%. People, however, are predisposed to focus on the negative things we hear and see, which now saturate us constantly through media and Internet sources and are reinforced by 'groupthink'. Therefore, when asked about the world, we delve into our minds and find our preconceived, undesirable ideas, which are usually the problems that surround us.

We certainly need to be careful about facts and the reliability and validity of evidence, especially from the Internet, which needs tighter regulation. Teaching children that much of what they hear and see may not be true, but a biased perspective, is a vital task. Students require judgement, clear thinking and confident communication to respond to events. This arises from inner language, which thinks, reflects and rationalises issues in our minds to then discuss with others and balance views. However, Hurlbert's team (2013) shows how only 20% of us are able to talk to ourselves in this way, which may connect with limited levels of narrative language and thinking being reported. Children need regular chances to orally express themselves *externally* in order to develop *internal* thinking, so opportunities at home and in primary schools are vital for facilitating this process. An economist colleague recently said that we would rather die than think, when discussing the Institute for Government's statement that our politicians show '*a lack of insight*' from the bad decisions made today! Promoting thinking through talk opportunities is what is necessary, but teachers have to be creative to enable this to happen in a prescriptive curriculum system.

Social media was designed to democratically empower people through self-publishing their ideas. This could then challenge the traditional hierarchies. Media giants would no longer have ideological monopolies and state secrecy would weaken. Anyone with a social media account has a public image to promote their interests. However, this vision of Utopia has soon turned to cynicism, with a reputation for incubating trolling, fake news and personal attacks. In reality the social industry was not invented to free us but to capture people's social life for profit. The result is increased anger, depression, self-harm and suicide. The social industry has a formula to control user engagement, reflecting wealth, status and competition. It eats time as smartphones and screens penetrate leisure hours, with the average person spending 135 minutes daily on social media platforms (Seymour, 2019).

> 'We are all slaving away as its unpaid "digital serfs"'
> 'Life on social media devolves into a war of all against all'
> 'The Internet has created new vectors of cyber war'
> 'The social industry politicises mental distress'
>
> (quotes from Seymour, pp. 22–25)

This tool is supposed to liberate and, in exchange for data, promises new ways to participate and speak up, but has intentions of which we are ignorant. Teachers have an important role in alerting students to such issues.

WHAT WE NEED TO LEARN

People require understanding of history, geography, psychology, anthropology, sociology, communication and creative arts, as well as mathematics, pure and natural sciences and computers. A broad knowledge and wide range of competencies are vital today with the need for flexibility in employment. People must be adaptable and comfortable enough to analyse chaotic, complex, open systems that are the order of today. Therefore, it is important to learn the practical and creative arts for constructing imaginative worlds and a capacity to invent. Teachers can help students to understand the intersection between the sciences and humanities. They should be encouraged to limit the tiny, closed worlds of gaming that they indulge in for fun, which can easily become addictive and time-consuming. Awareness of what helps or hinders balanced human development is important, with both teachers and parents making sure that children acquire this understanding. Learners need tasks that embrace all narrative thinking and language levels, to include the classroom ability range and enabling them to learn from observing each other's performances (see Sage 2000).

Conventional wisdom has been about a focus on narrow endeavours, but having a broad, generalist background with occasions to share views through face-to-face talk is what leads to brilliant breakthroughs (Epstein, 2019). In schools, colleges, universities and workplaces we must give learners opportunities to have a voice about what is important to them and encourage the things they appear to be good at. Students should regularly present their ideas, knowledge, skills and experiences to others so that they can build confidence, motivation and gain feedback from audiences. Learner participation is vital for successful futures, and the view that teachers are '*sages on stages*' must be replaced by '*guides by sides*' if real improvements in performance are to be made. Educators require a greater breadth and depth of knowledge and skill to facilitate such student learning effectively.

The Internet has given us the idea that there is a perfect universe out there, with means to achieve faultless living systems. However, it is a complex, muddled world of human beings with many different views, values, attitudes and customs that frequently clash. People often rub each other up the wrong way, resulting in grave misunderstandings that expand, explode and disrupt relationships. We skirt around life, making assumptions about the universe, but many of its secrets are undiscovered and will only be explored once we attain the '*know-how*'. The new industrial revolution is wrestling with ways to explore world mysteries and improve society.

THE FOURTH INDUSTRIAL REVOLUTION

The Fourth Industrial Revolution is interacting with other socio-economic and demographic factors to create a deluge of business-model changes, resulting in major disruptions to all labour markets. New categories of jobs will emerge, partly or wholly displacing others but creating new roles. Griffiths (2019) predicts the 7 top jobs of the future to be:

1. Algorithm Bias Auditor
2. Staff Empathy Consultant
3. Health Data Analyst
4. Automated Fleet Vehicle Scheduler
5. Voice UX Designer
6. Bot Manager
7. Robotic Licence Auditor and Registrar

While these new roles are difficult to grasp, all existing ones must already adapt to the presence of intelligent machines. Remarkably, 65% of children entering primary school today will end up working in completely new occupations that do not yet exist, according to the World Economic Forum's *The Future of Jobs Report* (2018). We need to make sure that learners can take knowledge and create new processes and ideas, as well as evaluate these against things like ethics and sustainability. No longer will a degree or diploma be enough to gain work. Higher education must be in the context of a greater range of knowledge as well as personal and practical competencies for flexibility in today's employment. Permanent jobs and 9–5 working patterns will become a rarity in an increasing gig economy. At one end is *'Finest'*, with gigs for only the brightest minds, whose performance and empathy metrics pass a high threshold. At the other end is *'Worka'*, where jobs will be available for miserable tasks, such as content moderation on social media (Griffiths, 2019). The reality may not be this *'black and white'*, as people often obtain employment because of their network connections.

The skill sets required in both old and new occupations will change in most industries and professions, transforming *how* and *where* people work. It may also affect female and male workers differently and transform the dynamics of the industry gender gap. *The Future of Jobs Report* unpacks and provides specific information on the relative magnitude of these trends, by industry and geography, with the expected time horizon for their impact to be felt on job functions, employment levels and skills. It is a report that all who prepare citizens for their futures should read, with a focus on learning *how* and not just *what* to learn.

THE AGENDA FOR THIS BOOK

This book unpicks the political, economic and social issues surrounding education that pose challenges in schools, colleges and universities, as well as workplaces. These suggest that a different approach to learning, assessing, teaching and training is needed to develop relevant experiences that allow all students to achieve their potential. Learners show a huge range of abilities and interests that they must share, with only a broad, varied, flexible curriculum satisfying their needs. In 2011, an Ofsted report, *Removing Barriers to Literacy,* said *'a common feature of the most successful schools is the attention they give to developing speaking and listening'* (p. 6). Heeding student voices and expecting them to pay attention to their teachers and each other is important in building respect for everyone, whatever their background. Sharing knowledge brings awareness of what you know and need to acquire.

Each chapter looks in detail at the many influences that there are on communication and language development and its use, to provide information for reflection relevant to anyone involved in teaching and learning at any level. This is the competence that experts regard as most essential for survival, which now requires greater levels of expertise. In Cuba, the view regarding education is that it is a *communicative process*, and unless one understands how it develops and breaks down it is impossible to teach effectively. Their educational performance is enviable, according to research and international reports. We need to give communication priority for standards to rise and make new inventions possible. The Japanese support this by putting *communication and relationships* at the top of their learning agenda. In some schools, students regularly teach their peers, as sharing what you know is the best way to learn, remember and apply knowledge practically to solve daily problems.

TEACHERS MUST TALK *WITH* STUDENTS AND NOT JUST *TO* THEM

Experience demonstrates that teachers are better employed as '*guides by sides*' rather than '*sages on stages*'. Although this facilitative approach happens in educational practice, this *guiding* approach needs to be implemented more regularly to help students achieve better personal competencies and master *how* rather than just *what* to learn. Chatting to new students entering senior school recently, I was told that their teachers talked for all of each lesson, which often bored them, before dishing out homework in the last five minutes. Where is the student voice here? Educators have to meet learning targets, which encourage transmission teaching, designed to acquire knowledge rather than personal and practical abilities. This method alone does not prepare students sufficiently for a future world where they need to connect well with colleagues and show initiative and independence. To run classes, where learners have greater control and input, so that they learn *how* to learn, requires more knowledge of educational arts and sciences than UK teachers acquire in a limited training. We must reflect on this, so that students can survive in a world demanding personal and practical abilities that traditionally have taken second place to academic, factual learning.

MAIN POINTS

- Evidence suggests that students lack *narrative language* for learning
- This deficiency may be contributing to a decline in *fluid intelligence*
- Teaching must become more three-dimensional, focusing equally on *academic, personal* and *practical* abilities to encourage creative and communicative independent learners

- This requires that students are given more *voice* in their educational experiences with some *control* over how they can learn best
- Tasks for students should reflect all narrative levels to include the whole ability range
- Learners need a strong *general background* and balance of *academic and practical* experiences to achieve *communication, collaboration and creativity* to solve problems

REFERENCES

The Basic Skill Agency: Skills for Life (2002) Report on Communication. www.skillsforlifenetwork.com/article/the-basic-skills-agency/2530 (accessed 12 Sept. 2019)

Cattell, R. (1963) *Theory of Fluid and Crystallized Intelligence: A Critical Experiment. Journal of Educational Psychology,* 54, pp. 1–22.

Demeneix, B. (2017) *The Toxic Cocktail: How Chemical Pollution Is Poisoning Our Brains.* Oxford: Oxford University Press

Epstein, D. (2019) *Why Generalists Triumph in a Specialised World.* London: Riverhead Books

Flynn, J. & Shayer, M. (2017) *IQ Decline and Piaget: Does the Rot Start at the Top?* www.gwern.net/docs/iq/2017-flynn.pdf

Griffiths, S. (2019) *7 Top Jobs for the Future. Engineering & Technology,* 14, October 2019

Hoare, J. (2019) *The Woman who Cracked the Anxiety Code.* London: Scribe

Hurlburt R., Heavey C. & Kelsey J. (2013). *Toward a Phenomenology of Inner Speaking. Consciousness and Cognition,* 22 (4), pp. 1477–94 PMID: 24184987

Mehrabian, A. (1971) *Silent Messages.* Belmont, California: Wadsworth

Ofcom Report. (2019) *Children and Parents: Media Use and Attitudes.* www.ofcom.org.uk/__data/assets/pdf_file/0034/93976/Children-Parents-Media-Use-Attitudes-Report-2016.pdf

Ofsted Report. (2011) *Removing Barriers to Literacy.* London: Ministry of Education

Sage, R. (2000) *Class Talk: Successful Learning through Effective Communication.* London: Bloomsbury

Seymour, R. (2019) *Willing Servants. New Humanist,* Autumn 2019. 134 (3). London: The Rationalist Association

World Economic Forum (2018). *The Future of Jobs Report.* weforum.org/reports/the-future-of-jobs-report-2018

SPEECHLESS

CHAPTER SUMMARIES

1. THE LEARNER CONTEXT

This chapter focuses on some issues that underpin learning problems. Educators are expected to teach to nationally required standards, with students tested at specific stages within the primary, secondary and tertiary systems. This one-size-fits-all philosophy ensures that transmission instruction reigns, with the teacher talking in long stretches to the class. Many learners do not prosper in a regime where they have to deal with large chunks of spoken and written language, because they have problems in assembling and expressing meaning. This situation is now common in a society relying on *text* rather than *talk* communication. Therefore, special needs and special educational needs are a growing problem in Britain, which does not respect the fact that humans develop at different rates and in ways not aligning with a prescriptive philosophy. The job of teachers is complicated further by continuing immigrant influxes. These students often lack the background to cope with the British system as their culture has followed a different child-rearing trajectory. Although the Department for Education dictates must be followed in schools, colleges and universities, being fully aware of the influences on teaching reinforces the importance of participatory approaches which are explored in later chapters of the book.

2. LEARNING IS COMMUNICATION: SIGNS, SIGNALS, SYMBOLS AND CODES

To teach people to communicate effectively for learning from others, one must understand the nature and structure of this complex process. The idea of codes lies at the heart of human communication and instruction, and without the use of signs, signals and symbols, which comprise this system, the exchange of messages would be impossible.

The chapter looks at these aspects of communicating in order to raise awareness of their importance in teaching and learning in plural societies where language differences can cause misunderstandings. It helps comprehension of the many different verbal and non-verbal aspects of communicating effectively with one another that take account of various intercultural customs.

3. ENGLISH AS THE INTERNATIONAL LANGUAGE

The English language dominates our planet and will continue to do so as the medium of the World Wide Web. Initially it spread across the globe with the march of the British Empire gaining and retaining prominence because of the economic success of English-speaking nations. English hit the million-word jackpot in 2009, with, appropriately, '*Web 2.0*', defining future web products and services. Of course, many of its words have origins in other languages, due to cultural mixing. Finalists meeting the 25,000 citations set by the Global Language Monitor, tracking usage trends, included Silicon Valley, India, China and Poland. The Polish word '*Bangsters*' (*people responsible for predatory practices*) combines *banker* with *gangster* and was coined for those triggering the trillions asset wipe-out and financial crisis. At its current rate, English generates a new word every 90 minutes, and is spoken by nearly a third of the world's 7 billion+ folk. Traditional language study and teaching has focused on its components – *sounds*, *words*, *sentences* and their *graphical representations* – but now broadens to consider the context in which it is used and the other systems involved in making meaning. In Britain, this emphasis from communication research has had limited impact on educational practice, although at the North Wales Language and Business School (Glain Orme) a successful course is based on this extended model. Language is the mode whereby people communicate, but is ironically the main means whereby they fail to do so. The chapter considers *international* problems posed by human languages and styles of communicating as well as *intra-national* issues for mutual understanding from the many linguistic varieties used in science, education, medicine, law, religion and mass communication. Solutions are suggested to bring new communicative insights.

The importance of learning other languages is mentioned, as in our plural societies many jobs are dependent on having multilingual competencies and therefore broader perspectives on the world. This allows appreciation of other cultures as well as relativising your own to see it objectively. It is a vital for making people more tolerant and open to the rest of the world. In terms of science, Anglophones are fortunate that the world writes in English, but in the humanities, this is not the case. If you want to be published in an Anglophone paper, you must conform to what they think humanities are about. In many domains, the English nation is not the best in the world. Since the British do not commonly read literature in other languages, their research is regarded generally as having limited application.

4. MULTIPLE LITERACIES IN A GLOBAL SOCIETY

This chapter discusses a broad-based conception of literacy, including all verbal and non-verbal ways of symbolising and representing meanings, so highlighting the importance of communication in personal and academic success. Over the past 100 years the world has moved from spoken to written communication, with a mass education system that regards reading and writing as the hallmark of achievement. Success in written language depends, however, on literate structures achieved in formal speaking, which assemble quantities of information for instructing, informing and explaining. Today, we spend more time looking at visual display screens than talking formally, and there is strong evidence that a lack of speaking opportunity affects verbal comprehension and expression and higher-level thinking. Human development indicates that formal talk structures must be achieved to shift primary language (*spoken*) into secondary modes (*written*). The fact that many people show difficulties in spoken and written communication suggests more awareness of these processes in teaching. This encourages us to look at a range of interdisciplinary perspectives to inform us how to facilitate learning, focusing on second language acquisition (SLA) for migrants in order to facilitate social inclusion and employment.

5. EDUCATING THE WHOLE BRAIN

The media joke about people with *'two brains'* refers to their super intelligence. The joke is also on *us* as we *all* have *two brains* – a *verbal* and *non-verbal* one – with two ways of thinking. However, we exist on less than half our brain power and this is unpacked, unravelled and understood to raise performance. On the *right* side, we have one way of knowing. We *'see' imaginary* things (*mind's eye*) or recall real ones. Imagine a favourite food – its colour, shape, taste and smell. We *'see'* how things exist in space, understand metaphors, dream, fill in information/opinion gaps in talk/text, combine ideas to make new ones and assemble meaning (*synthesise*). If a thing is too complex to speak about we gesture. Describe a spiral object without hands! Images (*'seeing' within*) are idiosyncratic, non-verbal thinking ways –

intuitively, holistically and metaphorically. This is the *'seeing/feeling'* brain, communicating with ourselves and understanding whole things/ events. The *left* side works oppositely. It analyses, abstracts, counts, marks time, plans and states logically, with *words* expressing *thoughts*. If apples are bigger than plums and plums bigger than cherries, we say that apples are bigger than cherries. This illustrates the *left* brain – analytic, sequential, symbolic, linear, objective and verbal. It is the *'saying/hearing'* brain communicating thoughts conventionally. This chapter shows how both brains complement one another suggesting that education favours *left* brain growth at the expense of the *right*. It leaves us full of facts but unable to apply them judiciously. Small changes in how we learn can produce big results. Using both brains effectively may shoot us up the evolutionary as well as the educational rankings for a double blessing!

6. BRIDGING EDUCATION AND WORK

There is a mismatch between school and work requirements and a sense that what we have been led to believe about education and qualifications is somehow false. Research has found an inverse correlation between qualifications and job performance. Education has been viewed as good for society, but tends to have a corrupting effect on genuine learning – concentrating on its products (*examination results*) at the expense of processes (*application of understanding*). Job recruiters look for psychological and social aptitudes that are not easily defined or measured. These develop from attention to individual needs, interests and key competencies which have been sidelined in pursuit of test results and school rankings. Many teachers, however, view their job as not just preparing students for qualifications as a means of livelihood, but giving them a broader view of what a fulfilling life looks like. This requires rejecting a narrow academic life-course as obligatory and inevitable in education! A focus on developing the attributes and virtues required for successful living and working are advocated for the new industrial world. Teachers play a vital role in nurturing these and must give them priority.

7. COMMUNICATION COMPETENCE: SOLUTIONS FROM RESEARCH

This section assembles issues that are communication problems and presents some solutions from research. Today, changes in living are faster than ever before, due to the advent of intelligent machines that are taking over mundane tasks and creating opportunities for new human roles. AI is being used to recruit staff, using verbal and non-verbal measures in mobile phone interviews. However, the curriculum has been narrowed over the last 30 years to elevate scores on national standard tests. These dominate teaching, leaving little room for anything else, although the independent education sector is moving to change this, by encouraging a broader curriculum that focuses on transferable abilities. In professional training, courses have increased preparation in subject matter and minimised personal, professional development. A political agenda has trumped best practice and led to a generation that is neither fully ready for life nor work challenges. The chapter summarises this discussion and poses solutions that are rooted in a more focused attention on communicative competencies to ensure improved confidence, coping, co-operative and collaborative attributes for the 21st-century citizen.

8. FINAL THOUGHTS: THE EPILOGUE

Whatever the future, experts agree that *language and culture* will become the most important unifying factors amongst people. Education will need to take more interest in linguistic and cultural issues that presently divide populations and cause tension, distrust and problems in understanding and learning effectively. Research at the University of Leicester showed how, in city schools, children all entered with language and cognition ages about two years below chronological ones (Sage, 2000). Two decades later the situation has not changed, with teachers reporting that this continues to be a major issue in formal learning. The basic obstacle to effective communication is the problem of meaning, which is especially important in cross-cultural interactions. This final reflection brings together the conditions for effective communication with models of intercultural competence and ways to promote this in education.

1

THE LEARNER CONTEXT

PREVIEW

This chapter focuses on issues that underpin learning problems. Educators must teach to nationally required standards, with students tested at specific stages within the primary, secondary and tertiary systems. This one-size-fits-all philosophy ensures that transmission instruction reigns, with the teacher talking in long stretches to the class. Many learners do not prosper in a regime where they have to deal with large chunks of spoken and written language, because they have problems in assembling and expressing meaning. This situation is now common in a society relying on text rather than talk communication. Therefore, special needs and special educational needs are a growing problem in Britain, which does not respect the fact that humans develop at different rates and in ways not aligning with a prescriptive philosophy. The job of teachers is complicated further by continuing immigrant influxes. These students often lack the background to cope with the British system as their culture has followed a different trajectory in child-rearing. ,Although the Department for Education dictates must be followed in schools and colleges, being fully aware of the influences on teaching reinforces the importance of participatory approaches for effective communication.

INTRODUCTION

Children find face-to-face conversation '*too much effort*', preferring to watch YouTube, an Ofcom report (2019) claims. The media watchdog says that 4-to-16-year-olds would rather watch screens alone than meet up with friends or pursue sports and hobbies, as these are more effort than they feel like expending. Are you speechless?

We should all be shocked, reflecting on the consequences for both the individual and society's intellectual and physical development as well as general well-being. Studies show that children without the conversation moves required to shift into extended talk lack thinking and narrative structures to easily acquire the secondary languages of literacy and numeracy or express feelings to gain needed support (Sage, 2000, 2000a, 2003, 2003a, 2004, 2004a, 2007, 2010). Furthermore, this language deficiency constrains the development of higher-level thinking and reflective action. In March 2019, the Minister of State for Apprenticeships and Skills, Anne Milton, announced that university graduates are *'immature, irresponsible and cannot communicate effectively'*, telling how business leaders are *'always moaning about people who come out of university'*, as they are not prepared or up to speed with the workplace environment requiring productive colleague engagement. Many assert that increased university student numbers have lowered quality.

This is a generalisation, of course, as some young people display outstanding personal, practical and academic competencies. However, limited understanding of *cause* and *effect* underlies the escalating numbers of serious crimes of 5-to-15-year-olds (*Magistrates in the Community project – MIC*). 85% of prisoners have been excluded from school during childhood and are illiterate in consequence, involving both speaking and writing. Recent figures show that perpetrators of knife crime are likely to also have been excluded from education for unacceptable behaviour. The numbers of school exclusions increased by 66% in 2016–19 according to Department for Education figures. Research in a Diagnostic Child Development Centre showed children referred for bad behaviour always displayed higher-level language delay/disorders, with consequent speaking, thinking, reading and writing problems (Sage, 2000).

HUMAN NATURE

Experts have long debated human nature and interactions, concluding that we have instincts that foster the greater good, but others that encourage self-interest and anti-social behaviour. The former needs continual encouragement and support. A healthy human mind depends on building communication for sharing views and feelings, to learn from others, develop balanced perspectives and foster respect. Humans are social creatures and their potent brains have developed, through the communicative process, to maintain relationships with others for individual and group progress. Creativity is collective, with talk and debate necessary for the collaborative spirit that brings about higher-level problem-solving (Haslam *et al.*, 2019). A collapse of community spirit recently, along with an erosion of personal and civic responsibility, is noted by experts and discussed in *Paradoxes in Education* and *The Robots Are Here* (Sage *et al.,* 2017, 2019). These books show how societal changes have strengthened individual rather than group interests, so lessening communication and cooperation to result in more personal stress. Brooks (2019) suggests that individualism has made us unhappy and the antidote is to communicate, connect and commit. The road to happiness lies in 'we' not 'me'.

A capitalist philosophy,* producing endless consumer goods, has encouraged personal wants above needs, to make us all more greedy and grasping as well as anxious to keep abreast of peers. In Britain, the welfare state and corporate bodies have replaced traditional community groups like friendly, mutual and religious societies. These institutions were based on reciprocity and nurtured trust and support, but have now been largely taken over by nationalised health and education services, corporate industries and government quangos. This has taken away joint responsibility, obligations, duty and pride – imposing obedience and engendering condescension and conflict instead. However, recent national financial problems are encouraging communal action again, like '*foodbanks*' to feed the homeless and voluntary car services in rural areas to help those without transport, since buses have stopped.

Nevertheless, the mandatory nature of the welfare state has brought about resentment in contributors and apathy in users, driving exploitation of the system for personal gain. Forceful government increases dependency, making people more selfish as well as less communicative and cooperative. Anthropological studies use the '*beehive*' metaphor to model effective communication, collaboration and harmonious living, advocating *small* working collectives, directed to the greater good (Axelrod, 1984). Exploring the logic of communication and cooperation, Axelrod showed that to build this '*one good deed deserves another*'. To nurture reciprocity, children must have the opportunity to develop effective informal and formal communication in a stable, repetitive context with a consistent approach. Operating in small groups, not larger than 150 persons, has

---

*Capitalist philosophy promotes private property, capital accumulation, wage labour, price systems and competition. Critics says it puts profit over social good – encouraging inequality, with power in the hands of a few, and corruption and economic instability. Others think it brings prosperity and increases production and ranges of goods and services.

shown throughout history to be necessary for harmonious living and working. Small units enable everyone to know each other, so regulating behaviour as social approval is necessary to survive happily. Crime and social conflict soar in places and institutions that are large in size, which illustrates the fact that small is beautiful and people pressure works. You cannot ignore or abuse people you frequently encounter.

What makes cooperation possible is the ability to exchange information and ideas and share views in a clear, structured, comprehensible way. We transmit messages verbally and non-verbally (*gestures, facial expressions, voice tone, manner, etc.*), with language reflecting the traditions, customs, knowledge and beliefs of users. This enables culture to be passed on from one to another and across generations. The existing 7,000 world languages and 20,000 plus dialects produce culturally distinct groups, with different values and customs that easily conflict when competing for power and advantage. Boyd and Richerson (1990) point out that to make communication and cooperation possible there has to be a degree of *conformity* when living and working in the large settlements that are the norm today, with the population tripling over the last century. These researchers reiterate: '*When in Rome do as the Romans do*', which is important for effective interpersonal communication, assimilation and integration. If you look, live and talk differently to indigenous neighbours, distrust and disrespect easily arises.

WHY IS IT VITAL TO FOCUS ON COMMUNICATION?

Regularly, the media relates stories about our population's declining language and communication, with a resulting dive in intelligence, confirmed in the Flynn studies (2013). Students find it difficult to read instructions on test papers and understand what the questions mean. At a recent seminar for business leaders, each speaker mentioned that communication and relationships were the greatest problems in improving employee performance, along with no initiative, which Sage (2003) attributes to lack of *inner language* to control thinking and action.

American scientists have developed a wristband that warns if you are boring people. This is needed because we are limited at picking up social cues now that face-to-face talk is decreasing in favour of communicating remotely by technology. Feedback from all sides of a conversation is analysed, showing how others respond to you. Also, the language and psychological traits of arguments are collected as data by these wristband sensors. The idea is that an app will act as a relationship counsellor, sending prompts to de-escalate tension if people have blazing rows. Machines are now telling us how to communicate and behave!

Smart machines are taking over many human activities and proving more reliable and cost-effective for routine tasks. This will free people to take on roles aimed at solving the problems of our increasingly complex, global world, but will require *creativity* and *innovation* – dependent on effective communication with oneself and others for cross-discipline collaboration (Fullan & Langworthy, 2016). However, Nicholas Carr (2010) in *What the Internet Is Doing to Our Brains* reminds us how technology has changed our concentration span and led to anger, division and hatred, encouraged by unregulated, often inaccurate web-based information that results in '*groupthink*'. Communication, through technology, produces disembodied messages devoid of feeling and humanity, and is now a common way to connect with people. This situation has encouraged cyberbullying, with English students suffering the most online attacks, according to the OECD Teaching and Learning International Survey of 15,000 schools (2019) out of 48 countries – discussed by Llewellyn. School leaders report that bullying has doubled since 2013. The i-SAFE Inc. survey (2019) suggests that 68% of students are victims of online bullying.

Such examples have led experts to suggest that education must focus on improved inter- and intra-personal communication for problem-solving, interaction and behaviour now that machines (*robots*) are taking over habitual tasks and requiring us to work smarter. At present, communication is an underdeveloped competence, witnessed when people shout at each other rather than conduct rational argument. Psycho-linguists are presently examining this discourse behaviour because of increased antagonistic exchanges in all walks of British life. Talking to *win* (objectivist) rather than talking to *learn* (relativist) means we are becoming less inclined to legitimise ideas that are different from ours (Fisher *et al.*, 2017). Future employment will rely on *interdisciplinary* communication and collaboration to jointly focus on solving present complex, political, economic and social issues, like people inequality. Teamwork is dream work, but needs effective cross-cultural communication for successful results (Ilieva, 2010).

WHY IS EDUCATION FAILING? ISSUES ABOUT CHANGE

Present demographic changes may seem unprecedented, but contact with other communities has occurred since the beginning of time. Change is a fact of living systems (*like societies*) arising from new influences (*forced or invited*), unforeseen events (*desirable or not*) and group efforts to maintain traditions and vary practice only within desired directions (Sage *et al.*, 2017).

Recent changes follow widespread transformations of cultural customs across human history. The development of farming and herding transformed cultural practices, beginning around 10,000 years ago in the East, moving from Turkey to Greece (8,500 years ago) and then throughout Europe. This enabled huge population increases, beyond which could be sustained by hunting and gathering (Diamond, 1992). The horse domestication about 5,000 years ago, amongst nomadic herdsmen from the Black Sea steppes, gave an advantage in travel and warfare, resulting in their language, culture and customs dominating continents. '*A few dozen horses helped Cortés and Pizarro, leading only a few hundred Spaniards each, to overthrow the two most populous and advanced New World states, the Aztec and Inca empires*' (Diamond, p. 240).

These sweeping changes, from herdsmen conquests, resulted in the takeover of many local languages throughout Europe and Asia by invaders. The herdsmen languages are the ancestors of the native tongues of about half the world's population. This is despite the fact that of around 7,000 languages only 140 belong to the Indo-European family. Expansion of European populations into North and South America has occurred over the last 500 years.

Ignatia Broker (Ojibway elder) reveals the massive changes felt by indigenous people worldwide and their effects and disruptions. The Ojibwe, Ojibwa, Chippewa, or Saulteaux, are an Anishinaabe people of Canada and the United States (US), who are numerous in the north of the Rio Grande. In Canada, they are the second largest First Nations population, surpassed only by the Cree. In the US they have the fifth largest population among Native American peoples, surpassed only by the Navajo, Cherokee, Choctaw and Sioux.

'*After the feast, in the evening, the people (of the Ojibway village) met in council to hear the news of the do-daim (visiting clansman). He told of a strange people whose skins were as pale as the winter white and whose eyes were blue or green or gray. "Are they good people or are they those who will be enemies?" asked A-bo-wi-ghi-shi-g. "Some are kind. Others speak good. Others smile when they think they are deceiving," replied the do-daim. "Many of the Ojibway have stayed with these people, but soon the people had great coughs and there were bumps on their skins and they were given water that made them forget". "Now," said the do-daim, "these strangers are many. They intend to stay, for they are building lodges and planting food... they cut trees and send them down the river... soon there will be no forests."*'

(Broker, 1983, pp. 18–24)

The result was that the village council sought a new place to live, because the ways of the colonising people were different and did not suit their traditional values and lifestyle, which had been developed over centuries to give them identity and stability. It shows how changes are sparked by contact with the outside world but deeply influenced and shaped by existing cultural practices in the community. Trade across distances reached all continents. Change intensified when monotheistic religions were established and colonisation, enslavement, cash, taxation, formal education, public administration and welfare services appeared in communities, as they became larger, more complex, sophisticated and multicultural.

Among the most significant are those involving work – especially for females. Traditionally, men have been hunters and women homemakers, but increasing society sophistication changed this, and consequently family relationships. Industrialisation required development of formal schooling to produce competent, compliant workers for rapidly expanding factories and offices. Male opportunities shifted outside their ancestral land, resulting in migration. Therefore, women took on more food-producing and caring tasks for families while husbands were away. This led to a decrease in women's influence in community political life, which was considerable in precolonial times. In Britain, the two world wars meant women were co-opted into male roles when the men went away to fight. Since many fighters were killed and did not return home, women continued in male jobs and now are career- as well as home-makers.

Changes in family and community structures are seen in parent-child relations. The logistics of education and work often separate families for long periods. This means shifts in child-care arrangements from kin-based to outside child-minding systems. Increasingly, in cities, child-carers have foreign backgrounds, with a non-English mother tongue. You see them in London parks with their child charges but conversing with other carers speaking their languages rather than talking English with the children. A headteacher 'blogger' on *Huffington Post* laments that many children enter school with inadequate language for learning, and lack of home talk a significant factor. Research at the University of Leicester (Sage, 2004–6) showed how all children entering a city school, over three years, had cognitive-linguistic ages at least two years below chronological ones, which doubled when in senior schools, with 80% functioning at a 5–6-year level at age 11–12. Outside child-minders limit parental involvement in the home as care-givers and contributors to the family. As the law has moved away from rights and obligations of lineages to those of individuals, parent powerlessness and frustration has emerged.

'Erosion of parental authority is further facilitated by the fact that nowadays children know more about contemporary life than do their parents. Consequently, most parents are finding it difficult to guide their children in how to behave in a world in which they are the more ignorant citizens.'

(Nsamenang, 1992, p. 137)

Nsamenang reported that changes and clashes between traditionalism and modernity have produced incompatible role demands and dilemmas that increase child psychological disturbances – now prevalent in education settings. Today's teenagers are almost twice as likely to be depressed as those 10 years ago, according to a University College London study comparing two cohorts (aged 14) born 1991– 92 (5,600 subjects) and 2001–2 (11,000 subjects) (Patalay & Gage, 2019). It also found that the recent cohort were more likely to self-harm, skip sleep and be obese, but less likely to indulge in drugs and alcohol, revealing a complex picture. Formal schooling has played a key role in this process. Although people globally look to education to improve their situation, the better jobs that this is expected to provide may not be forthcoming. *'Universities run threadbare degrees'* was a recent media headline, criticising a proliferation of low-value and low-quality courses that churn out graduates who cannot get jobs (Turner, 2019).

This point is reinforced by Glendinning *et al*. (2019) in the report: *Policies and Actions of Accreditation and Quality Assurance Bodies to Counter Corruption*. It suggests that unsuitable students, courses and assessment procedures may lead to cheating. In 2019, Pok Wong, who graduated with a first-class degree from a middle-ranking UK university, received an out-of-court settlement worth £61,000, claiming her qualification was *'Mickey Mouse'* and proved useless in securing a rewarding job with prospects. The Institute for Fiscal Studies has stated that those attending lower-ranking universities end up earning less than non-graduates. A drive for schooling may lead to success but also failure and loss. People entering Britain with another mother tongue not only lose this in their education instruction but may end up jobless. As more people seek higher levels of learning the number competing for suitable, available positions increases, leading to widespread alienation and unemployment. The huge increase in British university education means that many graduates cannot find work commensurate with degree expectations and suffer disillusionment that increases their mental-health issues.

Immigrants moving to Britain voluntarily in the hope of improving their economic well-being, political or religious freedom mostly do well in our fact-based education system. However, they are likely to have problems with the nuances of the English language, as well as vocal dynamics, because they inflect according to their mother tongue – making them difficult to understand. English is stress-timed, but many other languages are syllable-timed, and this causes misunderstandings (Sage, 2019). Comparing their current situation as better than in their old country, these people tend to see discrimination, language differences and other hardships as barriers to be overcome in learning English and a new culture. However, involuntary minorities are often disillusioned about their educational potential – regarding collective effort vital for achievement. Many end up in drug/sex trades as the only way to make money and attain a decent living. These minorities develop coping mechanisms to handle subordination in society, being regarded not just as different from the majority but in opposition to it. Boundary-maintaining structures, like living in cultural ghettos and forming gangs, help to retain identity and self-worth. Adopting majority ways would challenge cultural distinctions. This complicates learning of British customs and school engagement for minorities because of a resistance to compliance.

Schools may struggle to teach children from many backgrounds and abilities. In my first teaching post (1982), I had a class of thirty-six 14-year-olds to teach English. A third were born abroad and did not speak English at home and cognitive-linguistic ages varied from 4 to 18 years, with three students unable to write their names. This was before a National Curriculum and prescriptive teaching, so I could choose an oracy-to-literacy approach. Initially, I asked the class to select a poem and learn it to perform at the next lesson, when peers would assess to criteria agreed by everyone. This participant approach was successful and led to writing from group tasks giving opportunity for expression at all narrative levels. It enabled students to produce work at their development stage and so become successful learners. The pyramid below shows that learning by doing and teaching others is the best way to learn.

LEARNING STYLES PYRAMID
(National Teaching Laboratories, Bethel, Maine)
Percentage of learning retained

5% Lecture
10% Reading
20% Audio-visual
30% Demonstration
50% Discussion group
75% Practice by doing
90% Teach others/use immediately

GOALS OF DEVELOPMENT

Goals of development and what is regarded as mature or desirable varies amongst different cultural traditions and community circumstances. Popular theories specify a single developmental trajectory, resembling the values of the author's community and life course. Educated theorists regard literacy and Euro-American thinking as developmental goals. Lewis Henry Morgan (1877) proposed 7 stages of human progress: lower, middle and upper *savagery* and lower, middle and upper *barbarism* – to reach *civilisation*. Adams (1996) suggested that monogamy, nuclear families, *effective interaction*, agriculture and private property were basic to economic and social organisation leading to *civilisation*.

Recently, McGlone & Chatterjee (2019) have spoken about *touch* being the forgotten sense in healthy brain development for *effective interaction*. Rats whose mothers lick them regularly grow up better able to cope with stress, whilst those not receiving this become hypersensitive to anxiety. *Touch* slows heart rate, lowers blood pressure and reduces cortisol levels and feelings of social exclusion. Human interaction is being lost, but British teachers are forbidden to touch children. When a trainee teacher, I put my arms around a distressed child and was thoroughly reprimanded. In Japan, teachers often embrace students. When in Italian schools recently, children hugged their teacher when leaving the class at the end of a lesson.

Ideas of linear, cultural evolution emerged at the time when large bureaucratic structures were handling education in schools and economic activities in industry. Also, European influence was high in Africa, Asia and South America, with numerous immigrants from these nations seeking fortunes in the USA and Europe, because of their stable governments and economic successes. However, researchers have noted variations in the relevance of pre-literacy and literacy abilities in different cultures (Rogoff, 2003). In those where literacy is key to communication and economic success, preschoolers need to distinguish shapes, letters and colours, but if not central to community practice such activities are irrelevant.

In some cultures (*Japan, Cuba, USA and most European nations*), speaking is valued and the narrative style encouraged in schools, with literacy only introduced after formal, extended speaking has developed. I have seen in Italian, Japanese and Cuban schools children taught the different narrative levels to learn how to produce a *setting, characters/ things, events, actions, results* and *reactions* when speaking and writing. This does not happen to the same extent in Britain, as literacy and numeracy are introduced without checking that learners have the prerequisite literate speaking levels to support these secondary language modes. Many British teachers cannot name the *five conversational moves* needed for dealing with extended formal talk, because this knowledge has not figured in their training, but is deemed important information for nations valuing literate speaking (*follow the thread of conversation, ask and answer questions, assemble and contribute ideas and demonstrate maintenance behaviour – eye contact/smiling/ nodding, etc*.). In line with this, these cultures also respect brain development, with the right hemisphere responsible for the *overview*, dominant until 7–9 years, when the left kicks in to deal with *detail*. This is the age when children in the top 10 countries (educationally speaking) start formal school and attempt reading. In Britain, children are introduced to detailed sound analysis immediately, between 4 and 5 years, when the left hemisphere is not yet fully functioning to cope easily with this task (Sage, 2000, 2003, 2004).

Thus, in Britain we have many students classified as having *special needs* and *special educational needs* and regarded as failing in the system. Our racetrack approach assumes that children passing milestones of infancy and childhood earlier will be more successful adults. Classification of learners as '*ahead*' and '*behind*' is the achievement idiom. This is fostered by a prescriptive national curriculum with all students expected to follow a narrow path which values academic intelligences above personal and practical ones. Children who do badly in standard tests are deemed failures. In other parts of Europe and the world, personal and practical talents are valued as much as academic ones. In the Czech Republic, I found the status of a builder is the same as a brain surgeon, which is unlikely to be the British view.

An example of goal differences comes from West African mothers immigrating to Paris. They criticised the French use of toys, to get infants learning, as tiring for them. Some criticism related to a concern that an object focus may lead to impoverished communication and isolation, just as we might express the negative impact of screens on child development. African mothers prioritise social intelligence over technological, and often structure child interaction around other people, whereas the French focus on inanimate objects (Rabain Jamin, 1994). When interactions focus on objects, Africans stress social functions of these, like enhancement of relationships through sharing, rather than their use and action possibilities. Many cultures prize social solidarity above individual accomplishment. This is so in Japan. In art classes, I saw student groups producing *one* picture, which would not be witnessed in Britain, but was encouraged by the Japanese, rather than individual activity for higher engagement and performance.

With assessment finding that the Japanese outperformed British at all levels, the approach has merit (Sage, Rogers & Cwenar, 2006). The importance of family and neighbours has prevailed for generations, fostering the community above the individual.

Although cultural variation in developmental goals must be acknowledged, it does not mean that each community has a unique set of attitudes, values and direction, as there are some constants throughout these variations. Goals emerge from a specific way of viewing childhood as life preparation. This differs in communities where children participate in local events and are not segregated from adult life in preparatory settings like schools. To understand other groups we must look beyond established assumptions. Views reflect our particular cultural experiences. Accepting other ideas is uncomfortable knowledge, because some regard respect for other ways as criticism of their own. A learning attitude that considers where our views, values and customs are rooted brings understanding of human development, with universal features influenced by local variations. Communication, with insiders and outsiders of particular groups, encourages a balance and acceptance of human nature and how it functions.

Issues of our plural world are confusing today and land in the laps of teachers to unravel and bring understanding for learners. Added to the mix is a rise of transgenderism, as well as trans-racialism (*darkening and lightening skins*) and trans-ageism (*assuming a different birth age*) appearing on the scene. Psychiatrists work with biologically black children born to white parents, or adopted by them, pointing out problems faced, like sadness and confusion related to their identity. Understanding our origins and being proud of them is vital for self-awareness, esteem and well-being.

Knowing the background of learners is important for teachers. A teacher recently pointed out that in his school (*3,000 pupils*), with around 300 different languages spoken by learners, there was no time to investigate individual circumstances. This is the reality, but awareness of issues that underpin education helps reflection on how to implement successful learning. When children fail to learn within the system, the tendency is to focus on the attributes of individuals and give them arbitrary intelligence tests, which are restricted to only some of the factors that affect achievement. *Relational analytics*, used in therapeutics, focuses on the *relationships* with other learners, to plot inhibiting and facilitating factors. An example is found in *Class Talk* (Sage, 2000), looking at a student called Tom, in five different communicative classroom situations, to assess in what circumstances he learnt best.

REVIEW

This chapter focused on some issues that underpin learning problems. Educators are expected to teach to nationally required standards, with students tested at specific ages within the primary, secondary and tertiary systems. This one-size-fits-all philosophy ensures that transmission teaching reigns, with the teacher talking in long stretches to the class. Many learners do not prosper in a regime where they have to deal with large chunks of spoken and written language, because they have problems in assembling meaning from underdeveloped narrative language levels. This is exacerbated by the fact that children are now born with high cortisol levels, reflecting environmental stresses, which slow hearing and speech development (Sage, 2000). Research has shown that 75% of listeners have difficulties with this process, and only 12% of any lesson or lecture is remembered (Sage, 2003). This situation is now common in a society relying on *text* rather than *talk* communication and including many mother tongues. Conversations are *shallow*, about surface issues like holiday plans, rather than *deep* talk about what life means and how we feel about things. It is one reason why students suffer stress and anxiety that hinders progress, as they never get to grips with issues.

Therefore, special educational needs are important in British culture, which does not respect that human beings develop at different rates and in ways that do not align with a prescriptive philosophy. The job of teachers is complicated further by continuing influxes of immigrants, who often do not have the backgrounds to cope with the British system comfortably, as their culture has followed a different trajectory in child rearing. Figures show a 40% increase of immigrant students in British schools, which puts strain on scarce resources.

Although the Department for Education dictates must be followed in schools and colleges, being fully aware of the influences on teaching does reinforce the importance of participatory approaches. These are explored and explained in later chapters and seen in the *European Education for Robotics Curriculum*, which uses robots as tools for developing a holistic approach to learning to give personal, practical and academic abilities equal value. We must address the question of what our educational system requires to nurture intelligence about being a citizen in a plural society. This is ultimately understanding about language and behaviour, along with human collaboration in developing a shared world. A system obsessed with knowledge and its marketability and mainly interested in problem-solving educates only part of us. The shared human circumstance of living together, within principles of equality and well-being, needs the input of imagination, sympathy, faith and hope in all its forms.

Our present democratic deficit is that we are losing our idea of three-dimensional humans with personal, practical and intellectual competencies. With 50,000 thoughts a day, 95% of these are the same as before. It is estimated that when the Internet boomed, around 2003, humankind had generated 5 billion gigabytes of information. Now we create this amount of information every ten minutes, but our ability to deal with this remains as it was 70,000 years ago (Kenner & Rashid, 2018). A major international study, by a global team, reviewed world studies to assess the impact of the Internet on the human brain. It found three key areas where our increasingly digitised lives are changing the way we think and act, our ability to focus and concentrate, as well as memorise and socially behave (Firth & Sarris *et al.*, 2019). A decreasing cognitive and communicative capacity is seen world-wide. Learners must be helped to become smarter to deal with this phenomenon ,and this is the level at which we need to address our present crisis of confidence in the world. Drowning in information causes us all acute anxiety without strategies to deal with overload. University College London Neuroscience and Mental Health Group have just produced a study suggesting that 76% (50,000 subjects) find creative activities, like calligraphy and craft, help to block and reduce anxiety. It also finds 69% saying practical experiences boost confidence and 50% assert that they help thinking and reflection (Fancourt *et al.*, 2019). Such creative activities have taken a peripheral role in education but should have a more prominent place in attaining and maintaining well-being.

Dr Ian Pearson, a futurologist at Futurizon, was commissioned by AO.com (2019) to look at how robots are affecting life. He predicts that they will completely take over the average person's existence by 2030. Drones will clean homes, robot chefs prepare and serve our food, while we are kept warm by directed infra-red beams adjusted to our body temperatures. James Lovelock (2019) suggests that hyper-intelligent beings will inhabit the earth alongside us. They are carbon-based but not humanoid, and are probably spherical. These beings can communicate telepathically and instantaneously, perhaps using radiation to share their thoughts across vast distances. They harness quantum power so as to be in two places at once, or teleport between them. With the ability to think 10,000 times faster than humans, these new beings will coexist with their pitifully limited progenitors, curating the planet like a farmer tends the land in order to keep it fit for habitation. However, Lovelock's Gaia theory about the earth functioning as self-regulating means that humans will still be needed, just as we require plants and other animal species to survive for a balanced ecological system.

Human roles on earth, therefore, are likely to change rapidly, just as there has been a profound shift away from us shouldering responsibility in favour of demanding our rights. The individual comes before the group, with freedom to feel indignant when confronted with texts, speech and ideas deemed ableist, fattist, imperialist, nationalist, racist, sexist, transphobic, etc. Recently, the Army's recruitment campaign targeted '*snowflakes, phone zombies, binge gamers, selfie addicts and me, me, me millennials trapped in boring jobs*'. It was a controversial marketing approach, but it increased Army applications. The Infancy Studies Laboratory at Rutgers University is using brain-recording technology to understand the essential processes and changes occurring (like language) that underpin learning and behaviour (Choudhury & Benasich, 2011). Some countries (Japan and Cuba) make language and communication an educational priority, but in Britain there is no consistent policy, practice or training for teachers. At the 2019 Festival of Education at Wellington College, Amanda Spielman, Chief Inspector, lamented that debates at schools were being shut down by '*a worrying trend towards intolerance of different opinions*'. Clamping down on expression of views has huge consequence for higher-level thinking, balanced judgements and sharing thoughts for creative progress.

Preparing children for a dominant technology future is vital, and Europole has been operating an Education for Robotics in Europe for 10 years, monitored by universities. Students have a two-hour weekly programme, directed at developing personal, practical and academic competencies from building, coding and programming robots to acquire these abilities. Research (*in press*) indicates that this holistic, learning approach builds *resilience* that is vital for coping with constant changes and fast-paced living. In Britain, we are not proactive at implementing a mandatory, consistent, holistic approach to learning that will prepare the next generation to function well in the new industrial world. The situation is now urgent and this book addresses issues that need attention. We are existing in an uncertain present with the prospect of an alarming future unless we act now.

MAIN POINTS

- Studies support the fact that children enter school with insufficient narrative language levels for learning that require addressing for them to progress and reduce anxiety
- Schools now have many students whose mother tongue is not English and have to cope with issues of what is lost in translation in order for them to learn effectively
- Rapid society changes overwhelm, confuse and stress people, with prescriptive education meaning that greater numbers of learners now have educational needs

- Getting to grips with student backgrounds provides the information necessary to design effective learning that values a range of intelligences for overall well-being

REFERENCES

Adams, D. (1996) *Fundamental Considerations: The Deep Meaning of American Schooling, 1880–1900*. In E. Hollins (Ed.) *Transforming Curriculum for a Culturally Diverse Society* (pp. 27–57). Hillsdale, NJ: Erlbaum

Axelrod, R. (1984) *The Evolution of Cooperation*. New York: Basic Books

Boyd, R. & Richerson, R. (1990) *Culture and Cooperation*. In J. Mansbridge (Ed.) *Beyond Self-interest*. Chicago: University of Chicago Press.

Broker, I. (1983) *Night Flying Woman: An Ojibway Narrative*. St Paul, MN: Minnesota Historical Society Press

Brooks, D. (2019) *The Second Mountain: The Quest for the Moral Life*. London: Penguin Books

Carr, N. (2010) *What the Internet Is Doing to Our Brains: The Shallows*. London and New York: W.W. Norton & Company

Choudhury, N. & Benasich, A. (2011) *Maturation of Auditory Evoked Response from 6 to 48 Months: Prediction to 3 and 4 Year Language and Cognitive Abilities*. Clinical Neurophysiology, 122 (2), pp. 320–38.

Diamond, A. (1992) *The Third Chimpanzee*. New York: Harper Collins

Fancourt, D. (2019) *The Impact of Creative Activities on Anxiety*. London: NMH Group.

Firth, J. & Sarris, J. (2019) *The 'Online Brain': How the Internet may Be Changing Cognition*. World Psychiatry, May 2019

Fisher, M. *et al.* (2017) *The Influence of Social Interaction on Intuitions of Objectivity and Subjectivity*. Cognitive Science, 41 (4), pp. 1119–34

Flynn, J. (2013) *Intelligence and Human Progress: The Story of What Was Hidden in Our Genes*. Oxford: Academic Press

Fullan, M. & Langworthy, M. (2016) *A Rich Seam: How New Pedagogies Find Deep Learning*. London: Pearson

Glendinning, I., Maris-Orim, S. & King, A. (2019) *Policies and Action of Accreditation and Quality Assurance Bodies to Counter Corruption in Higher Education*. Washington, DC: The Council for Higher Education Accreditation

Haslam, S., Adarves-Yorno, I. & Postmes, T. (2019) *Creativity Is Collective. Scientific American*, Winter 2019

Kenner, S. & Rashid, I. (2018) *Offline: Free Your Mind from Smartphone and Social Media Stress*. London: Capstone

Ilieva, V. (2010) *Robotics in the Primary School*. Proceedings of SIMPAR, Darmstadt, Germany

i-SAFE Inc. (2019) *Report of Online Bullying*. auth.isafe.org/outreach/media/media_cyber_bullying

Llewellyn, C. (2019) *Cyber-bullying of Teachers: A Growing Problem for Schools. Teaching Times*

Lovelock, J. (2019) *Novacene*. London: Allen Lane

McGlone, F. & Chatterjee, R. (2019) *Touch the Forgotten Sense*. Podcast, January 2019

Milton, A. (2019) *Graduates not Ready for Work*. www.telegraph.co.uk/education/2019/03/08/graduates-immature-skills-minister-suggests-urges-youngsters/

Morgan, L.H. (1877) *Ancient Society*. London: MacMillan and Company

Nsamenang, A. (1992) *Human Development in Cultural Context: A Third-world Perspective*. Newbury Park, CA: Sage Publications

Ofcom Report (2019) *Children and Parents: Media Use and Attitudes*. Published by Ofcom Media

Patalay, P. & Gage, S. (2019) *Changes in Millennial Adolescent Mental Health and Health-Related Behaviour over 10 Years: A Population Cohort Comparison Study. International Journal of Epidemiology*. 27 February 2019

Pearson, I. (2019) *Robots Will Outnumber Humans by 2030*. www.futurizon.com

Rabain Jamin, J. (1994) *Language and Socialization in African Families Living in France*. In P. Greenfield & R. Cocking (Eds.) *Cross-Cultural Roots of Minority Child Development*. (pp. 147–66). Hillsdale, NJ: Erlbaum

Rogoff, B. (2003) *The Cultural Nature of Human Development*. Oxford: Oxford University Press

Sage, R. (2000) *Class Talk*. Stafford: Network Educational Press

Sage, R. (2000a) *The Communication Opportunity Group Scheme*. Leicester: University of Leicester

Sage, R. (2003) *Lend Us Your Ears: Understanding in the Classroom*. Stafford: Network Educational Press

Sage, R. (2003a) *Support for Learning*, Exeter: Learning Matters

Sage, R. (2004) *A World of Difference: Developing School Inclusion*. Stafford: Network Educational Press

Sage, R. (2004a) *Support for Learning*, 2nd Edition: Exeter: Learning Matters

Sage, R. (2004b): *Tackling Inclusion in Schools*. Victoria, Australia: Hawker Brownlow Education

Sage, R. (2006) *The Handbook of Social, Emotional and Behavioural Difficulties: Educational Engagement and Communication*. London: Continuum

Sage, R. (2006a) *Communication and Learning*. London: Sage Publications

Sage, R. (2006b) *Assessment and Teaching* the Communication Opportunity Group Scheme. Leicester: University of Leicester

Sage, R. (2007) *Inclusion in Schools: Making a Difference*. Stafford: Network Educational Press

Sage, R. (2007a) *COGS in the Classroom*. Leicester: University of Leicester

Sage, R. (2010) *Meeting the Needs of Students with Diverse Backgrounds*. Network Educational Press/Continuum

Sage, R. (2017) *Paradoxes in Education*. Rotterdam: SENSE International

Sage, R. & Matteucci, R. (Eds.) (2019) *The Robots are Here: Learning to Live with Them*. Buckingham: University of Buckingham Press

Sage, R., Rogers, J. & Cwenar, S. (2006). Report 2 of the Dialogue, Innovation, Assessment and Learning (DIAL) Project to develop the 21st century citizen. University of Leicester with Japanese Universities in Nara, Kyoto and Osako

Turner, C. (2019) *Threadbare Degrees Could Have Fees Cut. The Daily Telegraph*. 25 May2019

2

SIGNS, SIGNALS, SYMBOLS AND CODES: HOW WE COMMUNICATE

'Some hold translations not unlike to be the wrong side of a Turkey tapestry.'

James Howell, *Epistolae Ho-Elianae* (1645–55)
Book 1, Letter 6

PREVIEW

To teach people to communicate effectively for learning from others, one must understand the nature and structure of this complex process. The idea of codes lies at the heart of human communication and instruction, and without the use of signs, signals and symbols, which comprise this system, an exchange of messages would be impossible. The chapter considers these communicating aspects in order to raise awareness of their importance in teaching and learning in plural societies where language differences may cause misunderstandings. It helps comprehension of the different verbal and non-verbal aspects of communicating effectively with one another that take account of many intercultural customs.

INTRODUCTION

Learning Is Communication: Signs, Signals, Symbols and Codes
I am driving to the station and the ***30 mph sign flashes***. Oh no, I am going too fast – 32 mph pops up on a roadside screen! A sign with a

car outline shows the station car-park, and an *arrow* marks the driving direction. I stop at a machine, press a button and out pops a *barcoded card* with the *entry time*. Running to the ticket office, I find my credit card, greeting the manager: *'Hi, Bob, the sun's coming down in large drops as usual.'* I show my plastic card and key in my *PIN* while the scanner completes the transaction for my London fare. On the train, I skim a *crossword* before opening my laptop and typing in the *password*.

The narrative of my day in the big city could be continued, but boredom might set in rapidly! You will have noticed, however, the frequency with which codes enter our lives (*in bold italics above*). Also, unconsciously, you have been busy filling in the *information* and *opinion gaps* in order to get the gist. A 30-mph speed limit approaching the railway station is not made explicit, but the flashing light signals a warning which was learnt about in other contexts. Knowledge and experience allow us to understand correctly and behave appropriately in response to messages as long as we legitimise this information to influence actions.

Meaning is achieved through *codes*, which are systems of signs and symbols agreed between the communities using them. Systems are signs (*icons, indices, signals and symbols*) with rules for use, relating them to each other and aspects of reality. Codes represent reality to be understood, evaluated and sometimes changed in relation to needs. When travelling the world, we must learn the local codes for coping with our existence, otherwise there will be bother. Today, there are numerous codes operating to govern our working and private lives and this chapter aims for awareness of their significance and encourages us to take account of them in our teaching and training activities. Definitions of relevant terms that we use to transmit messages to one another begin our investigation.

SOME DEFINITIONS

Communication: Dance and Larson (1976) collected 126 definitions of two kinds. *Communication* is:

1. A *process* where a sender directs a message through a channel (medium) to a receiver with effect.
2. A *social activity* where people in a culture create and exchange meanings according to reality experienced. Messages cannot be transmitted without conversion into a code relevant for the channel used. Thoughts are encoded in *gestures, voice, facial expressions, manner, speech, writing and pictures*, etc.

Signals are *physical forms* for messages: *utterances, writing, gestures, telephone calls, radio transmissions*.

Signs have *physical forms* referring to something apart from themselves. *E.g. arrow → (direction)*.

Icons look like what they stand for, as in a *passport photo*.

Symbols have no visible connection with their signifier. *E.g. €, £, representing euro and sterling*.

Indices are unintentional signs like *footprints, smoke curls, wall cracks, etc*.

Encoder is a *physiological (brain)* or *technical (electricity media)* device transforming a message from a *source* into a *form (code)* for transmitting to a receiver to *decode*.

Decode is a *sense organ/receptor/technical device* converting a signal into a form for understanding.

Standard English is the norm for formal public speaking/writing – from *International Phonetics Association* rules

A WORLD SHAPED BY CODES

Codes are part of our world and history. Greek generals employed them to give secret orders in ancient times. Morse code, for signalling, is an international character encoding scheme used in telecommunication. It encodes text characters as standardised sequences of two different signal durations, called *dots* and *dashes* (*dits* and *dahs*) and is named after Samuel Morse, the telegraph inventor. The International Morse Code encodes the 26 English letters A–Z, some non-English ones, Arabic numerals and a set of punctuation and procedural signals (*pro-signs*). There is no distinction between upper- and lower-case letters. Each symbol is formed by a sequence of dots and dashes, with the dot duration the basic time measurement unit. The duration of a dash is three times that of a dot. Every dot or dash, within a character, is followed by a period of signal absence (*space*) equal to the dot duration. Words are separated by a duration space equal to three dots.

For efficiency, Morse code was designed so that the length of each symbol is approximately inverse to the frequency of occurrence in the English language character that it represents. Therefore, the most common English letter, E, has the shortest code – a single dot. As elements are indicated by proportion, not specific time durations, the code is usually transmitted at the highest rate that a receiver can decode. The transmission rate (*speed*) is in groups per minute. It is usually transmitted by on-off keying of an *information-carrying medium*, like electric current, radio or sound waves and visible light, which are present during the period of the dot or dash and absent between them. Morse code can be memorised and signalling perceptible to human senses (*sound waves/ visible light*)

International Morse Code

1. The length of a dot is one unit.
2. A dash is three units.
3. The space between parts of the same letter is one unit.
4. The space between letters is three units.
5. The space between words is seven units.

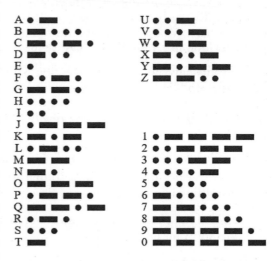

is interpreted by trained people. Many non-English languages do not use the 26 Roman letters, so Morse developed. The SOS distress signal – 3 dots, 3 dashes and 3 dots – is internationally recognised as the standard emergency signal and a Morse code pro-sign. In emergencies, Morse code is suitable for improvisation, like turning a light on and off, tapping an object or sounding a whistle, making it versatile telecommunication with complex thinking behind the code (Schneier, 2004).

A story of code-making is also that of code-breaking. Egyptian hieroglyphics (*sacred carvings*) became a coded system, as did the pictograms of the Chinese and Japanese languages.

智慧　原則

WISDOM　　PRINCIPLE

勇気　自由

Ancient tomb paintings show cattle-branding in place over 6,000 years ago. Today, to prove beast ownership, cowboys have a *branding alphabet* based on three elements: *letters or numbers*, *geometric shapes* and *pictorial symbols*. Originating in the 18th century, they were designed to identify cows over long distances and combat cattle-rustling, as shown here.

⌐ Broken bar		♡ Broken heart	
— Rail		♡ Flying heart	
/ Slash		⌐ Key	
⌒ Half circle		▦ Tumbling ladder	
○ Circle		⩊ Sunrise	

MODERN GLOBAL CODES

Communication technology (*computers and mobile phones*) has produced new sets of codes that save time and costs by making messages concise. These are modern forms of short-hand, but carry social kudos for teenagers! Examples are below:

Leet/Leetspeak
People world-wide are creating a new language code, which is the digital equivalent of Pig Latin (*see below*) with added hieroglyphs. Widely used in chat rooms, among hackers and gamers, leet (*corruption of 'elite'*) enables speedy communication to take place by using keyboard characters as short-cuts for sounds, words and phrases. Rules are:

- Letters are replaced by numbers resembling them, so 1337 = leet, 'B' ('be') is represented by '8', '9' replaces 'G', '5' stands for 'S' and '4' is 'A'. 13375p34k = leetspeak.
- Sounds are replaced by letters, so words ending in C or K now end in X, and Z supplies 'S' and 'ES' endings. However, you will find variations.

A leet rule is there are no conventions! Spelling and grammar rules are ignored and the idiom is constantly evolving (Churchouse, 2002). Here are some leet words and phrases:

Cool	k3w k3wL
Talent	m4d sk1llz (mud skills)
Hi	y0 ('yo' – alternative to 'hi')

Txt Tlk/Txt Lingo
Txt tlk is mobile phone language with similar aims to leet, but adapted for handsets and used in Internet chat-rooms. It developed from the Short Messaging Service (SMS), allowing text messages of up to 150 characters for conversations, news and financial information services. The first message was sent in 1992, and over the next decade the rate was 500 billion annually. Ofcom's 2012 *Communications Market Report* finds text messaging doubling annually and now the way to communicate. Text rules and examples are:

- Vowels are omitted
- Words are left out if sense not affected
- SpacesAreMarkedByCapitals

2	To or too
4	For, or as in b4
C	See
R	Are
Y	Why
8	Syllable 'ate' as in gr8 or m8
Bf or gf	Boyfriend or girlfriend
Thx	Thanks
Np	No problem

Keyboard or handset characters are also used to create pictorial messages like these ones: O:-) = Angel; :-! = Bored; %) = Confused (*tilt your head if you cannot see how this works*).

SPOKEN CODES

The simplest spoken code is Pig Latin – referred to above. It is a letter rearrangement code popular with youngsters. There are three basic rules:

- Words starting with vowels have '*ay*' added to the end, so '*actually*' becomes *actuallyay*
- Words starting with consonants have that letter moved to the end + '*ay*': '*can*' = *ancay*
- Two consonants at the start are moved to the end + '*ay*': '*speak*' = *eakspay*

'*A child can speak Pig Latin*' is said as: '*Aay ildchay ancay eakspay igpay Atinlay*'. A variation is Tut Latin, with the sound '*tut*' added between each syllable.

Opish is another spoken code in which '*op*' is added after each consonant. So '*code*' is '*copodop*'. Turkey Irish adds '*ab*' before each vowel, so '*code*' is now '*cabode*'. These are fun languages rather than used for meaningful communication. However, Cockney rhyming slang, spoken by indigenous East Londoners, has survived for over 200 years. It normally substitutes two words for one that rhymes with it. Traditional examples are:

Stairs	Apples and pears
Hair	Barnet fair
Dead	Brown bread
Shoes	Canoes
Shirt	Dickie dirt

Theories exist about how this language started in London's East End. It was devised to prevent other listeners from understanding what was being said and probably originated from:

1. Villains working in the docks
2. Market traders discussing customers
3. Prisoners passing information to keep it from guards
4. Thieves aware of police around

As a child, I remember going into an East End shoe shop to be told: *'Canoes – apples and pears.'* I was puzzled until my aunt whispered *'shoes upstairs'* in my ear!

More complex codes have been used to disguise information. Before the Germans occupied Norway in 1940, telephone and radio calls from Nazi agents were intercepted. On the surface, they were sending information about fishing, but were actually conveying details of ships, using numbers identifying vessels from Lloyd's Register. Later, in the Second World War, calls discussing the flower market disguised information about ships in harbour under repair.

CHANGE OVER TIME

On a bus, ending the London visit that began this chapter, I expected to relax from Queen's Park to Euston Station. However, this was interrupted when teenagers entered and spoke unrecognisably to each other. I heard: *'Ha, ha, ha'*, which I learnt meant *'three of us'*. They rang a mobile phone shouting: *'Hi bae, ITOY, Zip Ghost YOLO'* (*Hello babe, I was thinking of you high on marijuana and having a hard time – you only live once!'*). Another member of the group shrieked *'LMAO'* (*I'm laughing my ass off*). Why did English appear so strange to me? It was modified in a way that shocked and surprised me. I felt like a stranger – out of touch with loss of identity. English, however, is rarely heard on this bus, as other languages predominate.

English has suffered a metamorphosis. A standard language must prevail in public contexts to be widely understood – reflecting our roots with pride. Therefore, schools should focus on teaching students standard English to be understood in any situation. Other forms (dialects) must be valued and spoken when appropriate. Language, like all cultural aspects, evolves and adapts to user needs. If English had not changed over fifty years, we would not have words for modems, fax machines, cable TV, computers, iPads, iPhones, Internet, etc. In addition to new words, the evolution of slang phrases has a social function. Through talk, writing, music and art, different cultures communicate and express themselves creatively. These language modes evolve from exposure and experience to new people, places and things to incorporate different words, phrases, signs, symbols and signals as descriptors. English and its paralanguage today is different from that of Shakespeare, appearing as more functional.

'O, spirit of love, how quick and fresh art thou' Shakespeare, *Twelfth Night*, Act 1, Sc.1

Language reflects our changing culture and has evolved many times since its inception. There is Old English or Anglo-Saxon (*c*.450–1066 AD), Middle English (*c*.1066–1450 AD) and Modern English from Shakespeare onwards. Influences for change include:

- **Movement of people across the world** – *migration* and *colonisation* (e.g. *English speakers today are heard using the Spanish word 'loco' to describe a crazy person*).
- **Users of a language in contact with those speaking a different one** – no two persons speak the same: people from separate areas talk differently and within a community; variations are according to age, gender, ethnicity, social and educational background (e.g. *the word 'courting' has become 'dating'*).
- **New vocabulary for inventions** – like transport, domestic appliances and industrial equipment, or sporting, entertainment, cultural and leisure reasons (e.g. *19th-century term 'wireless' is today's 'radio'*).

Due to influences, a language embraces new words, expressions and pronunciations as people come across them daily and then integrate them into their own speech.

WHAT CHANGES HAS ENGLISH SEEN?

As English has changed, one can identify words now commonly used. Phrases like LOL (*Laugh Out Loud*), YOLO (*You Only Live Once*) and *'bae'* (*abbreviation of babe/baby*) have embedded over the past decade, as evidenced on the bus journey just described. While some words or abbreviations come from the Internet or text conversations, others appear as new words, a different meaning for an existing word, or one becoming more generalised than its former meaning, because of any one of the above reasons. Decades ago, *'blimey'* was a new expression of surprise, but now *'wow'* is more common usage.

Sentence structure has changed. Years ago, it was normal to ask, *'Have you a moment?'* Now, you might say *'Have you got a mo/sec?'*

Similarly, *'How do you do?'* has become *'How's it going?'* *'See you sometime'* is now *'See you later'*, which used to mean *'later that day'* but now refers to a longer time span. Sentences have been abbreviated, new responses given to everyday questions and words replaced with modern versions. *'Shall'* and *'ought'* are disappearing, but *'will'*, *'should'* and *'can'* are still in vogue, with other subtle changes. Some verbs take a complement with another in the *'ing'* or *'to'* form: *'They like swimming/to swim'*, *'I tried speaking/to speak'*. Both are used, but *'to'* has shifted to the *'-ing'* form (Sage, 2010).

WHAT DO CHANGES MEAN?

What other alterations to English have you noticed? Have they affected your teaching or learning methods? Most linguists accept that change in language, like society, is inevitable. Some think this is regrettable, but others recognise it reinvigorates language, allowing subtle expressive differences. Professor David Crystal (linguist) in a University of Buckingham lecture (2019), considered whether *'text speak'* undermines the English language. He pointed out that abbreviations have always been used. While some are new, others, like *'u'* for *'you'* and the number 8 as a syllable in *'later'*, have been around for over a century. Research shows a correlation between the ability to use abbreviations and spell, as you must know the letters for this. Language-user needs change and so will the language!

Standard English in Decline Among Teenagers
Half of teenagers fail to spot the difference between standard English grammar and colloquial language, according to research (Sage, 2018). Many GCSE English students do not grasp that phrases *'get off of'* and *'she was stood'* are grammatically incorrect. The use of social networking websites and mobile phones is thought to undermine literacy. Teachers complain that youngsters spend time watching screens instead of speaking, thinking and reading books. The result is that youngsters speak and think less and are unable to understand the consequences of their actions. The Crime & Safety Schools Initiative, part of the Magistrates in the Community (MIC) national project, came to Daventry schools (July 2019) with the police, prisons, fire service and judiciary running participatory sessions to show how a lack of thinking leads to actions with severe negative lifelong effects for themselves and others. The debrief session suggested that this annual event was the best day in the school year, helping students talk and think to benefit their behaviour.

In 2008, the Cambridge Assessment Examination Board surveyed 2,000+ teenagers in 26 English secondary schools. They were presented with phrases and asked to mark those employing non-standard English. Only 41% realised that an adjective had been used in place of an adverb in the phrase '*come quick*'. Fewer than 6 in 10 students could correctly distinguish between '*off*' and '*of*' and identify '*she was stood*' as ungrammatical. GCSE students (25%) failed to spot errors in the phrases '*it wasn't me who done it*', '*couldn't hardly move*', '*Tom had gotten cold*' and '*three mile*'. At least 20% failed to recognise that '*more easier*' was incorrect. Almost 1 in 10 students did not spot the use of a double negative in the phrase '*I didn't break no vase*'.

For many people their daily use of English is in new media, where non-standard grammatical constructions are acceptable. This inevitably leads to an increased lack of awareness of standard constructions. A Cambridge Assessment study (2005) found that GCSE students had a better grasp of grammar and punctuation than a decade ago but were more likely to lapse into colloquialisms. Non-standard forms will find a way into standard English, as teenagers are linguistic innovators. Every generation develops its own language based on contemporary youth culture, with social media and digital communication meaning teen slang evolves faster. It is difficult for parents to keep up with the latest youth words, like 143 (*I love you*), or PAW (*parents are watching*). In the appendix to this chapter is a list of 100 common acronyms in use. From social media acronyms to text message shorthand, most teens can converse with parents having no idea what they are talking about. It is important to be educated about slang to be aware of conversations online and in person.

GENERAL SLANG AND THAT DESCRIBING PEOPLE OR RELATIONSHIPS

Most teen slang words are used to sound '*cool*' or fit in with peers. Relationships are a vital aspect of adolescence, as teens learn who they are from interactions with one another. They create general words and those that describe friendships and romances, as seen below:

General	Describing People/Relationships
Awks – Awkward	**Bae** – '*Before anyone else*' – describes boyfriend or girlfriend
Cancel – A rejection of a person, place or thing	**BF/GF** – Boyfriend or girlfriend
Cheddar – Money	**BFF** – Best friends forever
Dope – Cool or awesome	**Bruh** – Same as '*bro*'
GOAT – Greatest of all time	**Creeper** – Someone socially awkward or with stalker tendencies

Gucci – Good or cool

Hundo P – 100% certain

Lit – Amazing

OMG – abbreviation for '*oh my gosh*'/'*oh my God*'

Rides – Sneakers or shoes

Salty – Bitter

Sic – Something that is cool

Skurt – Go away

Snatched – Looks good

Straight Fire – Hot or trendy

TBH – To be honest

Thirsty – Trying to get attention

V – very

YOLO – You only live once

Curve – Romantic rejection

Emo – Emotional or a drama queen

Fly – Boys refer to good-looking girls as 'fly'

Hater or h8er – Someone hating everything and everyone

noob – A person not wanting to learn

Ship – Short for '*relationship*'

Squad – A group of girls that hang out together

Tight – In a close relationship

Wanksta – A person acting tough but not pulling it off

COMPOUND SLANG AND IMPORTANT WORDS TO KNOW

Teens often combine two words together so one must know the definition of each word:

Chillaxin' **– Chilling and relaxing**
Crashy **– Crazy and trashy**
Hangry **– Hungry and angry**
Requestion **– Request and question**
Tope **– Tight and dope**

Bullying and cyber-bullying are serious issues today, so noting slang words which may indicate someone is bullying or a target for this behaviour is important:

Bye Felicia **– A disrespectful way to dismiss someone unimportant**
CD9 **– Code 9, parents are here**
POS **– Parents over shoulder (*texting to warn that mum/dad is reading*)**

Throw Shade – **Give someone a dirty look**
Tool – **Refers to someone stupid/geek**

Social media, text messaging and online dating can lead to sexual conversations. If not sure what to look out for, you might not notice what a youngster is saying. Talk of drinking or abusing drugs must be addressed immediately, with slang words – like CU46 (*see you for sex*).

GNOC – Get naked on camera
Molly – MDMA, a dangerous party drug
NIFOC – Naked in front of the computer
Netflix and Chill – a front for inviting someone to make out or
    more
Smash – Casual sex
Turnt Up – To be high/drunk
Zip Ghost – Someone high on marijuana and having a hard time
    functioning
Dexing – Abusing cough syrup
Crunk – Getting high and drunk at the same time
X – Ecstasy
53X – Sex
WTTP – Want to trade photos?
LMIRL – Let's meet in real life

Teen slang changes constantly, and the website *Urban Dictionary* updates this and is a resource for parents and teachers, but it is user-submitted and contains crude content. Phone apps help translate teen slang, such as the *Acronyms and Abbreviations Dictionary* and *Chat Slang Dictionary* apps.

COMMENT

The examples show codes are at the heart of human communication. Without a range of signs and an agreed way to use them, communication is impossible. How much communication could you convey without speech, body language, vocal expression, writing and other kinds of graphic representation? You could suggest *mind-reading*, but even here thoughts are encoded and decoded. Codes have arisen as humans have developed their communicative abilities of gestures, speech and writing (*in that order*), with each medium employing its own channel. Gestures use the *haptic* modality. Spoken languages, evolving over time, use the *acoustic* channel. Writing, which is relatively recent, depends on the *visual* mode for effect. However, what we regard as codes have been invented for specific purposes. Sign languages were introduced for hearing difficulties and Braille for the blind, using raised dots and the touch and movement channels. To solve communication over distance, Morse code was invented to transmit over electric telegraph wires at great speed. Semaphore and other flag signals convey information visually. All codes have features to distinguish them:

1. **More than one sign:** a beacon is a signal, not a code. However, smoke signals may be elements of a code. Thus, codes are a set of signs from which ones may be selected. The Roman alphabet is familiar to the British.
2. **Signs combine from rules:** syntagm describes any sign combination. PULL on a door is a syntagm of 4 signs from the Roman alphabet. '10' combines two signs from a values rule: 1 in first position means 10 and not just 1.
3. **Meanings depend on group use:** cultural differences mean people misunderstand signs. Europeans first visiting England may not know why people drive at 30 *miles per hour* (mph) not *kilometres per hour* (kmph).
4. **Codes meet social needs:** talking drums and smoke signals sent coded messages over distances faster than runners in old times. The Highway Code regulates traffic flow, updating to changing conditions. Library codes classify books for selection.

When developing awareness of codes around us we differentiate between *presentational* and *representational* types. Encoders use *presentational* codes to present themselves to other people (*like dress*) without transmitting a separate message. Appearing in a smart suit, not jeans, creates a specific image. You may have a meeting or interview and wish to give a professional, serious impression. Choice of clothes is the message, as *appearance* is a vital aspect of non-verbal communication. Guiraud (1975) classified the groups below:

LOGICAL CODES

Language codes
  a) Alphabetic
  b) Word substitutes (*picture writing*)
  c) Paralanguage (*intonation, gestures, facial expressions, etc.*)

Practical codes
  a) Military (*bugle calls*)
  b) Traffic control (*Highway Code*)

Knowledge codes
a) Scientific (*algebraic formulae, diagrams*)
b) Diacritics (*intonation mark system for speech transcription*)

Prediction codes
a) The Zodiac
b) Tarot cards

Aesthetic Art and Literature codes
a) Conventions of rhythm, rhyme, verse-form
b) Conventions of drawing (*perspective*)

Social codes
Protocols (*clothes for specific occasions*)
Rituals (*wedding, funeral or naming ceremonies*)
Fashions (*status symbols*)
Games (*structure, rules, vocabulary*)

Representational codes, in contrast, produce separate messages from the encoder, like a poem, Morse radio signal, photo strip story or line of Braille. A versatile code is the binary one: Yes/No; On/Off; 1/0; +/-, or any opposite pairs. Morse is binary dots and dashes and Braille uses embossed dots and flat positions in a matrix. Binary codes are employed for counting, replacing the denary 10-base number system, and used to communicate with computers, with the 1/0 signs converted into +/-, switching the electrical charges through the central processing unit. A computer does what it is told, so coded programmes enable the processor to carry out specific tasks. A high-level computer language is used to write programmes. These are limited English codes with word order, punctuation and spacing rules, using various languages appropriate for different applications. The central processor converts input into ASCII (American Standard Code for Information Interchange), representing pictures, sounds, letters, numbers and symbols (commas, brackets and carriage returns) as numbers. The numerical code can be decimal (denary), octal, hexadecimal or binary, according to the computer/level of processing.

This discussion shows how codes develop into sophisticated verbal and non-verbal representations over time. They are mainly built of icons, signals and indices with only rudimentary semantics (*meanings*) and syntax (*word arrangements*) but elaborate pragmatics (*use*). Verbal language has double articulation (*phonemes*, individual sounds, and *morphemes*, sound units) with highly developed semantics, syntax, pragmatics and complex coding systems. We need awareness of the specific systems of interaction used by different cultures. Their typical beliefs, values and behaviour are expressed and perpetuated through various codes. When people from different cultures meet there are many communication problems, which are fundamental if neither speaks the other's language. Even if they use a common language, problems arise because of different cultural assumptions made. How close would you stand to a chance acquaintance? This is about four feet away in England, but Indians and Arabs stand closer, which is uncomfortable and threatening for some people. Also, many gestures have meanings specific to cultures. Japanese people point to their nose for '*me*'. The temple screw is a gesture meaning '*crazy*' in England but '*intelligent*' in Holland. A hitchhiker's erect thumb might cadge a lift in parts of Europe but bring abuse in Israel, where it is an obscene gesture.

Subtle issues are seen when settlers speak the language of their new homeland. Problems with intonation, stress and accent emit wrong signals and impressions. Indian languages, like Urdu and Punjabi, have little stress carrying over to their articulation, where they fail to differentiate between '*this*' or '*last*' *time*. Asians find the high-pitch, stress-timed speech of English people too emotional and impolite and switch off listening (Argyle & Trower, 1979). Many mistakes are made because of misunderstanding and a lack of awareness of the signs, symbols and codes of different cultures and their communications. The process is difficult because people are complex accumulations of interests, needs, differences and competing intentions. However, communication teaching/training must account for three factors:

1. *One's own internal assumptions, sign, symbol and coding choices*
2. *Assumptions and symbolic interpretations by other person(s)*
3. *Verbal and non-verbal coding itself*

Ethics might make us think that most communicators do not intend to upset or harm others and act to fit societal values, norms and codes. However, language is power and used to abuse and frustrate others. Making meaning from signs, signals, symbols and codes underpinning events and behaviours is not easy. We depend on others speaking our language, referring to things experienced, whilst assembling communication in orderly, predictable ways. In another country, where your language is not spoken, there are problems translating unfamiliar words, and a lack of shared cultural experience to make sense of nuances and idioms in an exchange.

REVIEW

There are three principles of language and meaning to take into account. Firstly, people have similar meanings only to the extent of their similar experiences. Secondly, meanings are not fixed, and change with experience. Finally, people respond to events in relation to their *own* experiences. What we say about the world reports our particular way of viewing it. Seeing it on other terms (*empathy/sensitivity/communicative awareness*) is difficult. When talking about our familiar world, we end up speaking about ourselves. This is the trap of any teacher who must guard against being ensnared by what is believed to be the right and only approach to help others because of particular knowledge and experience. Since the 1970s, standard English has not been routinely taught in schools, because it is deemed elitist. Research at the University of Leicester (Sage, 2000) found that understanding teachers was the greatest problem for students, because information was often delivered in words and accents unfamiliar to them. Where I was born in Nottinghamshire, people speak of '*going on the gas*' (*walking on the pavement*) and '*making mash*' (*brewing tea*). Most languages have high (*standard*) and low (*dialect*) forms, with students taught the former for public discourse. In understanding educational failure, communication is a factor that receives limited attention in teacher-training programmes. Standard English has not been enforced over the last 50 years, but is necessary for some work roles, like the judiciary and military, so people face discrimination if they are unable to master this mode.

MAIN POINTS

- Continual movement of people brings them into contact with different verbal and non-verbal language modes, which must be learnt to communicate effectively
- Language systems are dynamic, resulting from changing needs
- The many dialectal forms of English make understanding difficult
- Teenagers use slang language to appear '*cool*', and it changes rapidly due to media influence
- Adults need awareness of teen slang to monitor an escalating drug, sex and bullying culture

All illustrations and graphics in this chapter were done by students at Nara University, Japan (Architectural Faculty).

REFERENCES

Argyle, M. and Trower, P. (1979) *Person to Person*. London: Harper and Row

Cambridge Assessment Studies (2005 & 2008) www.cambridge assessment.org.uk/our-research (accessed 21 June 2016)

Churchouse, R. (2002) *Codes and Ciphers*. Cambridge: Cambridge University Press

Dance, F. & Larson, C. (1976) *The Functions of Human Communication*. London: Hilt, Rinehart & Winston

Guiraud, P. (1975) *Semiology*. London: Routledge

Sage, R. (2000) *Class Talk*. London: Network Continuum

Sage, R. (2010) *Meeting the Needs of Students with Diverse Backgrounds*. London: Network Continuum

Sage, R. (2018) *One Funeral at a Time! Why We Hold onto Old Ideas*. *The Buckingham Journal of Language & Linguistics*, Vol. 10. pp. 53–63

Schneier, B. (2004) *Applied Cryptography*. London: Wiley Publishing

En.wikipedia.org/wiki/Leet (Alphabet of Leet symbols)

members.tripod.com/~mr_sedovy/colorado44.html (Cowboy branding alphabet)

www.cockneyrhymingslang.co.uk/cockney/ (Cockney rhyming slang dictionary)

www.altertons.com/com/index.htm (Cockney rhyming slang dictionary)

www.bbc.co.uk/history/ancient/egytians/decipherment_03.shtml (History of codes)

www.faqs.org/faqs/cryptography-faq/part01/index.html (Answers questions on cryptology)

APPENDIX (FROM A DICTIONARY COMPILED BY LEICESTER STUDENTS FOR A PROJECT)

1. **143** – I love you
2. **2DAY** – Today
3. **4EAE** – For ever and ever
4. **ADN** – Any day now
5. **AFAIK** – As far as I know
6. **AFK** – Away from keyboard
7. **ASL** – Age/sex/location
8. **ATM** – At the moment
9. **BFN** – Bye for now
10. **BOL** – Be on later
11. **BRB** – Be right back
12. **BTW** – By the way
13. **CTN** – Can't talk now
14. **DWBH** – Don't worry, be happy
15. **F2F or FTF** – Face to face

16. **FWB** – Friends with benefits
17. **FYEO** – For your eyes only
18. **GAL** – Get a life
19. **GB** – Goodbye
20. **GLHF** – Good luck, have fun
21. **GTG** – Got to go
22. **GYPO** – Get your pants off
23. **HAK** – Hugs and kisses
24. **HAND** – Have a nice day
25. **HTH** – Hope this helps/Happy to help
26. **HW** – Homework
27. **IDK** – I don't know
28. **IIRC** – If I remember correctly
29. **IKR** – I know, right?
30. **ILY or ILU** – I love you
31. **IM** – Instant message
32. **IMHO** – In my honest opinion/In my humble opinion
33. **IMO** – In my opinion
34. **IRL** – In real life
35. **IU2U** – It's up to you
36. **IWSN** – I want sex now
37. **IYKWIM** – If you know what I mean
38. **J4F** – Just for fun
39. **JIC** – Just in case
40. **JK** – Just kidding
41. **JSYK** – Just so you know
42. **KFY** – Kiss for you
43. **KPC** – Keeping parents clueless
44. **L8** – Late
45. **LMBO** – Laughing my butt off
46. **LMIRL** – Let's meet in real life
47. **LMK** – Let me know
48. **LOL** – Laugh out loud
49. **LSR** – Loser
50. **MIRL** – Meet in real life
51. **MOS** – Mom over shoulder
52. **NAGI** – Not a good idea
53. **NIFOC** – Nude in front of computer
54. **NM** – Never mind
55. **NMU** – Not much, you?
56. **NP** – No problem
57. **NTS** – Note to self
58. **OIC** – Oh I see

59. **OMG** – Oh my God
60. **ORLY** – Oh, really?
61. **OT** – Off topic
62. **OTP** – On the phone
63. **P911** – Parent alert
64. **PAW** – Parents are watching
65. **PCM** – Please call me
66. **PIR** – Parent in room
67. **PLS or PLZ** – Please
68. **POS** – Parents over shoulder
69. **PPL** – People
70. **PTB** – Please text back
71. **QQ** – Crying. This abbreviation produces an emoticon in text. It's often used sarcastically.
72. **RAK** – Random act of kindness
73. **RL** – Real life
74. **ROFL** – Rolling on the floor laughing
75. **RT** – Retweet
76. **RUOK** – Are you okay?
77. **SMH** – Shaking my head
78. **SOS** – Someone over shoulder
79. **SRSLY** – Seriously
80. **SSDD** – Same shit, different day
81. **SWAK** – Sealed with a kiss
82. **SWYP** – So, what's your problem?
83. **SYS** – See you soon
84. **TBC** – To be continued
85. **TDTM** – Talk dirty to me
86. **TIME** – Tears in my eyes
87. **TMI** – Too much information
88. **TMRW** – Tomorrow
89. **TTYL** – Talk to you later
90. **TY or TU** – Thank you
91. **VSF** – Very sad face
92. **WB** – Welcome back
93. **WTH** – What the heck?
94. **WTPA** – Where's the party at?
95. **WYCM** – Will you call me?
96. **YGM** – You've got mail
97. **YOLO** – You only live once
98. **YW** – You're welcome
99. **ZOMG** – Oh my God (sarcastic)

3

ENGLISH AS THE INTERNATIONAL LANGUAGE

PREVIEW

The English language dominates the planet as the medium of the World Wide Web. It spread globally with the march of the British Empire, gaining and retaining prominence because of the economic success of English-speaking nations. With double the Chinese vocabulary, English hit the million-word jackpot in 2009 with, appropriately, 'Web 2.0', defining future products and services. Finalists meeting the 25,000 citations set by the Global Language Monitor, tracking trends, included Silicon Valley, India, China and Poland. The Polish word 'bangsters' (people with predatory practices) combines banker and gangster and was coined for those triggering the trillions asset wipe-out and financial crisis. English generates a new word every 90 minutes, and is spoken by a third of the world's population. Traditional study and teaching has focused on language components – sounds, words, sentences and graphical representations – but now considers the context in which language is used and other systems making meaning. In Britain, this research emphasis has had limited impact on educational practice. Language is the mode whereby people communicate, but ironically the means whereby they fail to do so. International problems of languages and communicating styles, as well as mutual understanding from linguistic varieties, are examined. Since English is the international language, British citizens do not speak other languages to the extent that other nations do, so limiting knowledge of other cultures and their ways of thinking. In a plural world it also restricts working opportunities, as multilingual speakers will be favoured in many occupations.

INTRODUCTION

How many languages exist in the world? 20,000? 15,000? 10,000? Are you close? In the *Ethnologue* catalogue there are 6,909 distinct world languages with 230 spoken in Europe and over 2,197 in Asia (SIL International). The University of Maryland's *Langscape* project has interactive maps and linguistic data for 7,000 world languages. In Papua New Guinea (*less than 4 million population*) over 830 languages are found, making the average number of speakers around 4,500. All languages belong to 40–50 distinct families, which may change with scholarship as there is no present count with scientific status. The reason is not just that places like Papua New Guinea or the Amazon are unexplored, but enumerating languages is complex. Missionary organisations, like SIL International, lead the field in documenting languages for translating the Christian Bible, with sections translated into 2,508 forms by 2009.

LANGUAGE FAMILIES

A **family** is **a genetically related language group**, with the Indo-European (*to which English belongs*) the best known. This group is widely distributed and influential, so one might think that most languages belong to this family. About 200 Indo-European languages exist, but disregarding those whose genetic affiliation cannot be determined, there are at least 250 others. French governments have aimed to make France linguistically uniform, but discounting Breton (*Celtic*) and Allemannisch (*Germanic*), spoken in Alsace, and Basque, *Ethnologue* lists at least 10 distinct Romance languages, including Picard, Gascon, Provençal and others, in addition to 'Parisian French' (Chambers & Trudgill, 1998).

Multilingualism in North America is discussed in terms of English v. Spanish, or immigrant languages like Cantonese or Khmer, but the area spoke many languages before Europeans or Asians arrived. Originally, over 300 languages were spoken, with half dying out and only early word lists or limited grammatical and textual records available. This leaves about 165 of North America's indigenous languages spoken today. Beyond major ones of economic and political power, like English, Mandarin, Spanish and those with millions of speakers, we find many others from genetically distinct families, but these are disappearing.

Top 10 Languages by Number of *Native* Speakers and *Total* Number
According to the **number of native speakers only**, these are the most spoken languages, researched for the INTERMAR* European Commission project (2012–15) on language diversity and problems of communication.

1. Chinese

The *Ethnologue* estimates 1.2 billion native speakers and a billion speaking Mandarin – the most spoken world language. Chinese is a tonal language with thousands of logograms.

2. Spanish

Spanish tops English with about 400 million speakers. The politics of language and identity are disputed. Ask Catalan or Quechua speakers if Spanish is their tongue and the answer differs. It is the primary language of most of South and Central America, Spain and parts of the USA.

3. English

There are over 360 million native English speakers and half a billion using it as a second language, showing its success for business, travel and international relations. The ease with which it is learnt (*compared with Mandarin*) and the power of US culture means that English is synonymous with opportunity.

---

*2012–15: INTERMAR (Inter-comprehension in European Marine and Military Academies): *European Commission research into language diversity and educational implications in 23 countries.*

4. Hindi

India has 23 official languages, with Hindi/Urdu the chief among them. Whether this is one language, Hindustani, or two dialects, is debated. Spoken mainly in north India and parts of Pakistan, Hindi uses Devanagari script, while Urdu has Persian notation. Moves to have Hindi displace English in south Indian states as the official language have been resisted. Hindi gave us *shampoo* and *bungalow*, amongst other words.

5. Arabic

Arabic has about 250 million speakers, but like Chinese is effectively a number of languages, grouped conveniently as one. Modern Standard Arabic is primarily written, related to the Classical Arabic of the Quran. Spoken Arabic in Oman and Morocco are so different that their people might be able to discuss ancient texts but struggle to order food in a restaurant.

6. Portuguese

From the 15th century, Portuguese traders and conquerors brought their language to Africa, Asia and the Americas. The spread of Portuguese was initially tied to European colonisation, but these countries developed their own vibrant language brand. Today, Portuguese is spoken by 215 million in Brazil, Goa, Angola, Mozambique, Cape Verde, Guinea-Bissau, São Tomé, Príncipe, Macau, Machado de Assis, Bossa Nova, Mia Couto, Fernando Pessoa and Agualusa.

7. Bengali

The 1947 Partition of Bengal by the British divided West Bengal (*mainly Hindu*), now part of India, from East Bengal (*mainly*

Muslim), now Bangladesh. It is the language of Kolkata, the Andaman Islands and over 170 million Bangladeshis. By the next century, the population is projected to double.

8. Russian

With over 170 million native speakers, Russian is the 8th most spoken world language. Known for its grammatical structure and Cyrillic script, it is one of six languages spoken in the United Nations.

9. Japanese

Most of the 130 million Japanese speakers live in Japan, which makes it the most densely concentrated of the 10 languages. It has two writing systems, *hiragana* and *katakana*, plus Chinese *Kanji* characters. The largest groups of Japanese speakers outside Japan are in the US, the Philippines and Brazil.

10. Punjabi

Punjabi has about 100 million native speakers in India and Pakistan. The Punjab was divided by the British.

The 10 top languages can be arranged according to the **total number of speakers**, which changes the order. English edges out Chinese for top spot. Japanese and Punjabi lose positions, with French and Indonesian joining, as more people speak them as a second language.

1. English: 121 billion	2. Chinese: 107 billion	3. Hindi: 534.2 m	4. Spanish: 512.9 m	5. French: 284.9 m
6. Arabic: 273.9 m	7. Russian: 265 m	8. Bengali: 261.8 m	9. Portuguese: 236.5 m	10. Indonesian: 198.4 m

Linguistic diversity is declining, as advancing major languages dominate. When a language ceases to be learned by children, it is unlikely to survive current speakers. The North American situation is typical. Of about 165 indigenous languages, only eight are used, by 10,000 people. Around 75 are spoken only by older people and are likely to disappear, due to pressure from European settlement. A quarter of languages have fewer than 1,000 remaining speakers, and linguists agree that at least 3,000 of the 6,909 languages listed by *Ethnologue* will vanish this century. This affects more world languages than biological species. Hebrew is a language that died but revived. When it ceased to be the principal Jewish language, it had become solely a subject of expert study, but then it became the official language of Israel. A group abandons a native language for one more economically advantageous. Where there is no dominant language and diverse heritages reside, multilingualism is natural. Once a language goes, a community's connection with its history, culture and knowledge is lost.

What makes languages distinct is social and political rather than linguistic issues, and number estimates are opinion, not science. Max Weinreich (1980) said '*a language is a dialect with an army and a navy*', when discussing Yiddish, which is considered a dialect because it is not identified with a political body. The distinction is implicit regarding European languages v. African dialects. What counts as a language and not a dialect involves matters of state, economics, literary customs, writing systems and aspects of power, authority and culture. Chinese dialects (*Cantonese, Hakka and Shanghainese*) are as different from each other (*and from Mandarin*) as Romance languages like French, Italian, Romanian and Spanish. They are not mutually intelligible, but status derives from association with a single nation, a shared writing system and government policy. In contrast, Hindi and Urdu are essentially the same system, referred to originally as Hindustani, but associated with different countries (*India, Pakistan*), writing systems and religious stances. Although Indian and Pakistani varieties used by educated speakers are more distinct than local vernaculars, the differences are minimal – less significant than those separating Mandarin from Cantonese. Serbo-Croatian was spoken throughout most of former Yugoslavia as a single language with different local dialects and writing systems. Within this territory, Serbs (*mostly Orthodox*) use a Cyrillic alphabet, while Croats (*mostly Roman Catholic*) use the Latin one. After Yugoslavia's political breakup, three new languages (*Serbian, Croatian and Bosnian*) emerged, although linguistic facts had not changed.

CAN MUTUAL INTELLIGIBILITY IDENTIFY DIFFERENT LANGUAGES?

Mutual intelligibility is if speakers of **A** understand those of **B** easily. **A** speakers can understand **B**, but not vice versa. Bulgarians consider Macedonian a Bulgarian dialect, but Macedonians say it is a distinct language. In 1995, when Macedonia's president, Gligorov, visited his Bulgarian counterpart, Zhelev, he had an interpreter. However, Zhelev claimed he could understand everything said by Gligorov. Kalabari and Nembe are Nigerian languages. The Nembe understand Kalabari without difficulty, but the more prosperous Kalabari regard the Nembe as poor relations whose speech is unintelligible due to different accents. Accent is the way you sound when speaking and is how others judge personal background.

There are two accents. One is *'foreign'*, when speaking a language using the rules or sounds of another. Sometimes, when having trouble pronouncing other languages, substitutes similar to mother tongue are used, which appear wrong to native speakers. Once I went to a meeting where all the doctors spoke English as an additional language, and I had trouble understanding them, as their accents were unfamiliar. The other accent is the way a group speaks their native language, determined by locality and social network. People in close contact share a way of speaking (*accent*) differing from other talk. You notice someone with an American accent if not from the USA as it is different to your speech. Everybody has an accent in another person's view.

We are born to perceive and produce all human sounds. By a year old, children have learned to ignore distinctions not in their mother tongue. With age, it becomes harder to learn the sounds of a new language. German speakers of English have trouble with the sounds beginning the words <u>wish</u> and <u>this</u>, as they do not exist in their language, substituting V and Z, the most similar alternatives in their system. The German *schön* (*beautiful*) and *müd* (*tired*) have vowels missing in English (*speakers learning German pronounce them with an English accent*). Japanese have trouble with English L and R, because their language does not distinguish them, so cannot differentiate *light* from *right*. An English speaker has similar issues with the Thai language's *'aspirated'* (*puff of air*) and *'unaspirated'* P. Hold a hand in front of your mouth and say *pot* and *spot*. You feel a puff of air against your hand when saying *p* in *pot* but not in *spot*. P aspirates at the beginning of a word like *pot*, but not after S. It is difficult to differentiate *pot*, with and without a puff of air. Thai aspirated and unaspirated Ps are as different as English L and R. Using one word for the other alters meaning (*'forest'* to *'split'*), just as L with R in English (*lap* to *rap*). English has different rules for aspirated and unaspirated P use, so learning Thai means making mistakes like the Japanese do with English L and R sounds (Anderson, 2002).

Sound patterns vary across languages. An English syllable may begin or end with a consonant cluster, as in ***strengths***. In Japanese, a syllable contains only one consonant followed by one vowel (***ma***, ***ki***) except when ending with N (***san***). Pronouncing English clusters is hard for Japanese speakers and they tend to produce a vowel between consonants. Native speakers master some sounds before others. In English **p, m, n, h** and **w** are first acquired, while **z, j, v** and the two ***th*** sounds (***think*** and ***the***) are mastered last. All sounds are generally learnt before puberty. But non-natives will that have long-term difficulty with sounds not in their own native language. Some languages have trilled Rs – '*clicks*' made with the tongue as air inhales, or sounds made further back in the throat than English. Also sounds similar to native ones cause '*interference*' when speaking new languages.

Sentence structure varies between languages too. In Russian there is no connecting word like *is* in: *The tree is tall*. A Russian learning English saying – *tree tall* – sounds foreign. Two languages may allow the same form but in different situations. English can have a word/phrase beginning a sentence: (1) '*Pasta I like but potatoes I hate*'. This word order exists in Yiddish but used in a *range* of contexts: e.g. (2) '*My friend likes spending; a car he's buying now*'. '*He's buying a car now*' becomes '*a car he's buying now*', just as in (1) '*I like pasta*' becomes '*pasta I like*'. Both (1) and (2) sound fine in Yiddish, but only (1) does in English, as in (2) the speaker displays a non-English sentence structure.

Children mostly acquire their mother tongue effortlessly, but learning another language later in life is a different matter. Some languages have more complex word-building rules and others intricate sound patterns or sentence structures. Despite differences, researchers have not found any one language or group to be more difficult in all respects. Difficulty in learning a language depends on its commonality with those already spoken. French, Italian, Spanish and Portuguese descend from Latin, so a speaker of one learns the others easily. Likewise, English, Dutch and German are related to an earlier Germanic form. Languages relating to a native one may bring problems, because similarity results in interference causing mistakes (Baker, 2001). A different language, like Chinese, Turkish or Mohawk, brings added issues. In Chinese, a '*word*' has consonants, vowels and also '*tone*' (pitch). The syllable *ma* uttered with high tone (*mother*) has a different meaning from *ma* with low-rising tone (*hemp*) and high-falling tone (*scold*), so difficult for the English to master. No language/group is harder than others, and all are easy for infants. Only those speaking something else find it difficult (Romaine, 2000). Unfortunately, in the UK the population's knowledge of other languages has declined, due to the fact that English is presently considered as the one of Internet usage. This means that monolinguals have limited opportunity to learn the culture and thinking patterns of other languages, which makes them less employable in a global world and perhaps not so able to appreciate the perspectives of others.

Why mutual intelligibility criteria fail to identify distinct languages is because of **dialect continua**. If you walked from Paris to Berlin, the people who provided you with breakfast in the morning would understand (*and be understood by*) those serving you dinner in the evening. The French speakers at the beginning and the German speakers at the end would have trouble, because they speak two distinct languages. In the Australian Western Desert this continuum is over 1,000 miles, with regional speakers understanding one another while the ends have problems. How many languages are represented? We refer to languages of Chaucer (1400), Shakespeare (1600), Jefferson (1800) and Trump (2000) as English, but they are not mutually intelligible. Shakespeare might have been able to converse with Chaucer, but Jefferson and Trump would need interpreters. Languages change, maintaining intelligibility across adjacent generations, but eventually developing different systems. Language distinctness is unresolvable. Political and social issues outplay linguistic reality, and mutual intelligibility criteria are ultimately inadequate.

Analysing external language facts gives limited insight (Bybee *et al.*, 1994). Speakers produce and understand what they say, hear or read, so **internalised knowledge of language grammar** is more useful, so we might count the range of distinct ones. What differentiates them? Aspects, like how pronouns are interpreted, seem common across languages. In '*He thinks that Luke is cool*', the pronoun <u>he</u> can refer to any male with one exception – <u>he</u> cannot be the same person as <u>Luke</u>. This is a fact about languages in general when structural relations are the same. However, the fact that adjectives precede nouns in English (a *blue ball*, not a *ball blue*) is opposite in French. With a parameter inventory, we could say that each collection of values, identified in the knowledge of a set of speakers, is a distinct language.

Consider negative sentence construction in northern Italy. Standard Italian uses a negative marker before verbs: *Riccarda non mangia la carne = Riccarda not eats the meat*, while the language of Piedmont has a negative marker after the verb: *Riccarda a mangia nen la carn = Riccarda she eats not the meat*. Other differences derive from this: standard Italian cannot have a negative with an imperative verb, using the infinitive instead, while Piedmontese allows this. Standard Italian has a '*double negative*': *Non ho visto nessuno = not have I seen nobody*, but not in Piedmontese. Negation here establishes a parameter to distinguish these and other grammars. Others distinguish one grammar from place to place, independently of those in northern Italy. Suppose 10 such parameters distinguish one grammar from another. If each varies independently of others, collectively they define a set of two to the tenth, or 1,024 distinct grammars, with experts estimating 300–500 of these possibilities in the region. All Italian forms have commonalities and ways in which they are distinct as a group from other global languages. Since possible grammatical systems expand exponentially as parameters grow, if only 30 the number of languages becomes huge – over a billion on this assumption. Not all possibilities materialise, but if the space of possible grammars is covered uniformly, as in northern Italy, for the limited parameters there, the number of world languages must be greater than the *Ethnologue's* 6,909 estimate (INTERMAR, 2014).

World languages seem diverse, but communication systems are similar. Human languages differ from other known living creatures, providing a non-stimulus-bound system with endless possible messages. This is achieved with finite units, combining hierarchically and recursively into larger ones. Words are structured from a small sound inventory – individually meaningless elements integrated systematically, independent of their sentence combinations. Principles governing sound systems, words and meanings are mostly common across languages, with only limited difference possibilities. Human language differs from other systems, as the narrow ways in which one grammar diverges from another is insignificant. For a Milan native, variances between their speech and that of Romans looms large, but for visitors both are Italian. Global differences in grammars seem important, but for observers are minor human variations (Anderson, 2002, 2012). These are linguistic issues, but in legitimising messages psychological ones come into play – discussed in the next section.

COMMUNICATIVE INTELLIGENCE

Talk as a spontaneous exchange of words, or planned presentation to learners, can alter a life permanently. Phrases like: '*You can't do that!*' and '*You're not up to it*' are quickly uttered, but can devastate. Communications trigger emotions and are rarely neutral, carrying history and baggage. Each experience overlays another within the brain and is activated in exchanges, conveying meaning more embedded in listeners than speakers. We connect with others formally and informally, and these acts bind us together. Words are not external reality, but project inner experience. Language is instinctual and utilitarian, signalling how to *explore, navigate* and *survive* situations. *Communicative intelligence* enables successful interaction with others. Through dialogues and monologues we build trust, bond and grow, developing partnerships to create and transform society. Communicative wisdom is our most powerful hardwired ability. Teachers must build sensitivity for words and their performance. Words either cause us to bond and trust and think of others as friends, or break rapport so that we regard them as enemies. Creating effective organisations through talk is vital for open minds and successful results.

CONVERSATIONAL LEADERSHIP (HARVARD BUSINESS SCHOOL)

Globalisation, new technologies and changes in how organisations interact reduce efficacy of top-down practice. What replaces this? The answer is in how leaders manage communication and the information flow to, from and amongst people. A dynamic *conversational* process has been explored at Harvard. Employer efforts to *talk with* people and advance this in workplaces produced a model: '*organisational conversation*'. Smart leaders engage in *person-to-person conversation* rather than issue commands. They initiate practices and foster conversational sensibility, allowing a large organisation to function as a small one. This retains or recaptures operational flexibility, engagement and strategic alignment, targeting interpersonal attributes: *intimacy, trust, listening* and *connectivity* for performance.

Intimacy
Conversation flourishes when participants stay close. Organisational conversation requires leaders to minimise distances – institutional, attitudinal and sometimes spatial. Where conversational intimacy happens, leaders seek and earn trust and attention. They cultivate listening and speaking directly, understanding '*phatics*' – the small talk forming connection with others. Physical closeness between leaders and others is not always possible, but mental and emotional proximity is vital. Adept leaders step down from on high and step up to communication challenges. Intimacy distinguishes organisational conversation from standard forms, shifting focus from a top-down information distribution to a bottom-up exchange of ideas. It is less about issuing orders and more about questioning, gaining trust, listening and making personal connections.

Trust
Where there is no trust there is no intimacy, with the reverse also true. No one will exchange views with someone who has a hidden agenda or

hostile manner. Discussion is rewarding only when each person takes the other at face value. Trust is hard to achieve, and leaders earn it by treating all equally and addressing topics that feel off limits.

Listening
Leaders taking conversation seriously know when to stop talking and start listening. Attentiveness signals respect for all ranks and roles. Listening sessions encourage participants to raise issues, with information gleaned that may have escaped attention.

Connectivity
Invite people to raise concerns and feedback on performance, with anonymous results appearing on screen for audience awareness of their impact, encouraging questions on behaviour. True listening takes *bad* with *good* – legitimising criticism. A principal introduced *communication for diversity*, using videos of leaders speaking about what this meant. A working-class teacher described a school where staff were middle-class: '*My accent was different. I wasn't included and felt stupid.*' Such stories make impressions. I have used an approach in staff training which asked participants to say something positive about colleagues.

INTERACTION

Conversation exchanges are comments and questions between two or more people (*dialogue*). One person talking is a *monologue*. This applies to organisational conversation, when teachers talk *with* and not just *to* learners, making communication open and fluid, not closed and directive. Reducing *monologue* and embracing *dialogue* vitality builds intimacy. Efforts to close gaps between teachers and learners fails without having means and support to speak up and talk back for understanding. Communication must be promoted and facilitated. For decades technology made it difficult to support interaction in organisations. The media achieved scale and efficiency with print and broadcast operating in one direction only, but new channels disrupt this structure, bringing the style and spirit of personal conversation. Interactivity is not just deploying the right technology but investing social media with *thinking*. Tradition works against transforming communication two-way but interactive cultures value a space for dialogue. TelePresence (Cisco) simulates an in-person meeting, beaming video feeds between locations. Multiple screens create a wraparound effect and specially designed tables mirror one another so users feel together, mastering visual scale. Studies of remote interactions discovered that if an on-screen image is less than 80% of true size, those viewing are less inclined to talk.

Participants, therefore, are life-size, and look one another in the eye, offering real conversation. A regular video blog, a brief, improvised email message, allows direct, informal speaking, suggesting immediacy, to build trust and encourage comments.

INCLUSION

At its best, personal conversation is an equal-opportunity experience with participants sharing ownership, putting hearts and minds into discussion to generate content. Inclusion adds criticality to intimacy and interactivity. Whereas intimacy involves leaders connecting with colleagues, inclusion focuses on the role played in the process. Employees provide ideas rather than defend those of others. They become ambassadors, thought leaders and storytellers to boost reputation. Inclusion means that leaders cede control over how things are represented, but cultural, technological changes have eroded this anyway. Self-regulation tempers top-down control: an unacceptable statement means others respond and feelings are balanced.

INTENTION

A rewarding conversation is open but not aimless, with participants knowing what they hope to achieve. When absent of intent, it meanders or shifts into a blind alley. Intent confers order and meaning to digressive chat. Over time, contributing voices must converge on a single vision of what communication is *for*, with a shared agenda to align with objectives. While intimacy, interaction and inclusion produce ideas, intention closes the process. It enables relevant action from ongoing debate. Leaders must convey principles, not by asserting but explaining them, generating rather than commanding assent and speaking explicitly about the vision and reason for decisions. Employees then gain an overview of where their workplace stands in its context. To help them understand strategy, they should partake in its creation. Most important is reviewing the positivity of one's views for cross-cultural interactions. Self-report on these in hypothetical situations demonstrates:

- **Tolerance of ambiguity** – *maintaining composure and well-being in uncertainty*
- **Positive cultural orientation** – *evaluation of cross-cultural situations as positive*
- **Cross-cultural self-efficacy** – *confidence to engage in cross-cultural situations*

Manner

- **Self-awareness** – *understanding impact of one's own culture, values, preferences*
- **Social monitoring** – *awareness of physical, verbal and non-verbal behaviours and cues of others*
- **Suspending judgement/perspective** – *considering other views and holding back one's own*
- **Cultural knowledge application** – *utilising relevant cultural knowledge in interaction*

Method

- **Behaviour regulation** – *monitoring and revising own behaviour to be culturally appropriate*
- **Emotion regulation** – *monitoring and revising emotions in a controlled way*

Communication is a basic competence to be continually developed. Even a natural communicator must seek opportunities to enhance abilities. Effective communicators develop empathy and trust with others – adapting their style to suit audiences and situations. There are four recognised styles describing how you generally act. Do you usually…

Take an active role in communication?
Take time to think and respond logically?
Connect, observe and empathise with others, treating them as
    equals?
Review a situation for effective responses?

Identify your style below (*you may combine styles*).

Initiator
When talking I am involved and miss other reactions.
I express myself clearly.
I interrupt a speaker if disagreeing with them.
I talk more than I listen and make my views heard.
If not engaged I end or divert discussion.
My attention wanders if bored.
I like to set the topic for discussion.
I can talk about a topic while doing something else.

Thinker
I deal with confusion or conflict up front.
I speak and write to the point.

If interrupted I lose the train of thought.
I am annoyed if others stray from the point.
I select the best way to convey messages.
I like to keep to a discussion agenda.
I make lists of what I need to say or do.
I deal with conflict constructively.

Observer

Shifting across topics does not concern me.
I repeat statements to check understanding.
I observe the body language of listeners.
I encourage contributions by asking questions.
I like to listen to and observe others.
I alter my language/pace to ensure others understand.
I draft and redraft written communication for clarity.
I value other viewpoints.

Reviewer

I review the best way to present ideas for effective responses.
I focus on correct information.
I use all channels to reinforce ideas (*speech/pictures/movements*).
I structure ideas appropriately for audiences.
I match verbal and non-verbal information.
If I don't understand I will figure it out later rather than speak up.
I like information that provides resolution to issues.
I do not like dealing with emotional people.

Having determined your communication style, ask:

> *Does this match communication work style? If not, what can
> make the styles compatible?*
> *Is this the most effective style for your role? If not, which style
> best suits?*
> *Will current competencies enhance your career? If not, how can
> you develop them?*

The more you communicate in situations the greater your ability to
deal with any environment, regardless of context or circumstances.
Whether speaking formally or informally, addressing an audience or
writing a report – basic principles are:

Know your audience

Communication should be packaged to suit the listener's understanding
level.

Know your purpose and topic
Clarify when giving or requesting information or being social. Be aware of facts and details.

Anticipate objections and present a complete picture
Objections arise from misunderstandings. Communicate benefits. Support with evidence.

Communicate a little at a time then check listener understanding
Pause, ask questions and give the listener an opportunity to clarify.

Present information in several ways
What works for one listener/reader may not work for another – use different information processes.

Develop practical, useful ways to gain feedback
Feedback is how to evaluate effectiveness of communication.

In life we play different communication roles, instigating an exchange or being its recipient. Personality is a set of characteristics determining how we think and act, like relating to others. How we think, what we feel and our behaviour characterise what others expect of us. You are known for being on time but suddenly become late. This conflicts with your *conscientious* reputation. Others might think something is wrong – or if they don't have this characteristic it may not be noteworthy. Personality affects interaction to impact on life. Sutin *et al.* (2009) found that neuroticism (*negative states*) had the greatest effect on success. Those with positive personalities tend to be rewarded in life and work. Although there is debate about whether or not personalities are inherent when born (*nature*) versus the way they develop (*nurture*), most experts agree that they result from both experiences. She acts like Mum – because born with maternal traits, as well as learning them whilst growing. What 10 attitudes would you select to represent your personality? How might they affect your actions?

A will to perform	Dynamism	Integrity	Respect for others
Accomplishment	Ease of use	Intelligence	Responsiveness
Accountability	Efficiency	Intensity	Results-oriented
Accuracy	Enjoyment	Justice	Rule of law
Achievement	Equality	Kindness	Safety
Adventure	Excellence	Knowledge	Satisfying others

All for one/one for all	Fairness	Leadership	Security
Beauty	Faith	Love, romance	Self-giving
Calm, quietude, peace	Faithfulness	Loyalty	Self-reliance
Challenge	Family	Maximum utilisation	Self-thinking
Change	Family feeling	Meaning	Sensitivity
Charity	Flair	Merit	Service
Cleanliness, order	Freedom, liberty	Money	Simplicity
Collaboration	Friendship	Oneness	Skill
Commitment	Fun	Openness	Solving problems
Communication	Generosity	Other's point of view, inputs	Speed
Community	Gentleness	Patriotism	Spirit, spirituality
Competence	Global view	Peace, nonviolence	Stability
Competition	Goodness	Perfection	Standardisation
Concern for others	Goodwill	Perseverance	Status
Connection	Gratitude	Personal growth	Strength
Content over form	Happiness	Pleasure	Systemisation
Cooperation	Hard work	Power	Teamwork
Coordination	Harmony	Practicality	Timeliness
Creativity	Health	Preservation	Tolerance
Customer satisfaction	Honour	Privacy	Tradition
Decisiveness	Human-centred	Progress	Tranquil
Delight of being, joy	Improvement	Prosperity, wealth	Trust
Democracy	Idealistic	Punctuality	Truth
Determination	Independence	Quality of work	Unity
Discipline	Individuality	Regularity	Variety
Discovery	Inner peace, calm, quietude	Reliability	Well-being
Diversity	Innovation	Resourcefulness	Wisdom

Source: www.gurusoftware.com/GuruNet/Personal/Topics/Values.htm
(permission for INTERMAR Project)

COMMUNICATION AND NARRATIVE THINKING

Brigman *et al.* (1999) argued that communicative competence predicts success, identifying:

Learning abilities – *to listen, attend, follow directions and problem-solve*

Social abilities – *to work and play cooperatively, forming and maintaining relationships*

These abilities are needed to overcome challenges when adjusting from *informal* home to *formal* school relationships (Hargreaves, 1997, Sage, 2000). Mills *et al.* (1998) studied education in Hungary, Switzerland and Belgium, finding they focus on *engagement* by teaching use of *eye contact, listening* and *memory*. Non-verbal communication is targeted (*voice pitch, pace, pause, power, pronunciation, gestures, manner, presentation, props, context, etc.*), with social learning stressing appropriate group behaviour, *spoken* language, phonological abilities and mathematical understanding of size, time, space and quantity. This '*message-orientated communication*' (Butzkamm, 2000) brings understanding of communication use. Mehrabian (1971) showed how affective message impact owes only 7% to words but 38% to voice and 55% to gestures, with paralanguage and context vital for meaning.

Is communicative competence an issue?
Studies have long raised concerns about communicative competence on school entry (Bell, 1991, Boyer, 1992, Lees *et al.*, 2001, Sage, 2002, 2004, 2017). Surveys identify communication as weak in workforces (Confederation of British Industry). Initiatives attempt to develop key abilities, like communication for 16–19 learners in England (Hayward, 2004). Discussing communication '*standards*' is fruitless, as comparing over time is problematic. However, the combined decline of manufacturing and growth of services has shifted workforce profiles. 'Blue-collar' jobs, demanding physical and craft skills, have lessened, while 'white-collar' work, needing interpersonal communication, has increased (Hayward ibid.). The implication for education, whether considered narrowly for serving the economy, or broadly for personal fulfilment, is that abilities to cooperate, attend, listen and express ideas creatively, critically and coherently are vital for success. *The Communication Opportunity Group Scheme* (COGS, Sage, 2000, 2017) emerged to develop these competencies.

What is the Communication Opportunity Group Scheme (COGS)?

The COGS facilitates communicative competence and narrative thinking response to research that underachieving learners had difficulties for two reasons (Sage, 2000a, 2000b, 2002, 2004). Firstly, they were expected to perform at a higher level in literacy than achieved orally. Secondly, they failed to comprehend extended talk and express sequenced ideas. The COGS facilitates these verbally and non-verbally by developing narrative thinking and communicative competence. Goals cover ability and age range, using the zone of *'proximal'* (Vygotsky, 1962) or *'potential'* development (Alexander, 2000). 5 tasks at each goal level: 4 oral and 1 written, at the same narrative level, are developed through *'message-orientated communication'* through aspects referred to as CLARITY, CONTENT, CONVENTION and CONDUCT (Table 1).

Table 1 – Description of Non-Verbal Aspects of 'Message-Orientated Communication'

Non-verbal	DESCRIPTION Sage, 2000b
Clarity	Making messages clear with word choices and through voice and gesture.
Content	Organising message topic according to purpose by making connections within it.
Convention	Awareness of protocol governing, giving and receiving messages in different contexts, as well as the way words form sentences.
Conduct	The image a person presents to others, shown by personal responses in communication – including eye contact and facial expression.

What is the role of narrative?

'Narrative' does not just refer to literary text, but includes oral narratives relating personal histories, justifications and future plans, as well as those experienced in life (Wood, 1998). Ricour (Frid, 2000) and Bruner (1990) view narrative as the 'organising principle' of experience, knowledge and transactions. In social constructionism, narrative thinking and communicative competence connect, stressing interaction and culture in shaping development. Interactions are 'formative', as talking to and with others exposes us to communicative functions. Listening to and telling narratives effectively requires a listener perspective and planning coherent, comprehensible speech sequences. Frazier (2001) suggests that gaps in meaning or action recruit the listener/reader to fill them. COGS has information and opinion gap activities, enabling participants to contribute and transform ideas into their own. Table 2 shows seven goals.

Table 2 – Progression of Ideas Embedded in the COGS Framework

Goal	Idea development	Description Sage (2000b)
1	Record	produce a range of ideas
2	Recite	order simple ideas
3	Refer	compare ideas
4	Replay	sequence ideas in time
5	Recount	explain ideas – why? How?
6	Report	introduce, discuss, describe, evaluate ideas
7	Relate	setting, events, actions, results, reactions

How is the COGS taught?

COGS is taught through tasks targeting specific and core communication competencies. Circle games create relaxation to aid interaction. A tell, show, do and coach approach includes systematic sequencing of teaching with review, demonstration, guided practice and corrective, supportive feedback as effective (Wang *et al.*, 1994). Five learning principles are:

- interactions are 'message-orientated';
- the teacher facilitates 'message-orientated' activities, modelling communicative compctencies;
- learning is active;
- activities are learner-centred;
- empathy and cooperation helped by valuing feelings and mutual learning support and self and peer assessment.

To share meanings there is group and independent activity, enacting Vygotsky's (ibid.) idea that what a child does co-operatively today they can do alone tomorrow. This is compatible with '*sustained shared thinking*' in *Effective Provision of Preschool Education* (EPPE), with trained professionals providing models and open questioning (Sylva *et al.*, 2004). COGS is compatible with *dialogic teaching* (Alexander, 2003). Firstly, it is collective, with teacher and pupils learning together. Secondly, it is reciprocal, as both listen to each other. Thirdly, it is cumulative, building and chaining ideas coherently. Finally, the atmosphere is supportive of free speaking. Participants are shown how to say things and what is expected, while encouraged to talk, move around and have fun. Taking risks with ideas is encouraged, as tasks do not involve '*right*' or '*wrong*' answers. This enables empathy and cooperation, as they express what they feel. Formative assessment involves teachers and peers telling participants what they do well, for positive feedback. At Goal 1, '*message-orientated*' activities are supported by games. To help develop non-verbal aspects of *clarity* a poem is performed.

Algy met a bear. The bear met Algy. The bear was bulgy. The bulge was Algy!

Sentences are simple, but the *message* must be inferred (*the bear ate Algy*), encouraging participants to fill the *information gap* for meaning. The aim is to practise speaking clearly by making words louder or slower and consider whether others hear and think about what is said. Developing *content* centres on learner interests, constructing sentences about themselves and families. *Information* and *opinion gap* games support this, like putting objects in a box, passed around to music. Participants take turns to pick one out, when music stops and talk about it. They are encouraged to bring their own objects to motivate talk. Activities for *convention* include practice sentences with correct word order, in meaningful contexts, asking questions and requesting politely. Tasks for *conduct* involve thinking about body language, like eye contact, when communicating and how questions are answered to give a good/bad impression. This supports Sylva's (ibid.) findings that effective settings encourage children to rationalise and talk through conflicts.

THE STUDY CONTEXT (EXAMPLE OF A SMALL STUDY WITH ITS RATIONALE)

Research took place in a primary school. Children came from an estate with long-term unemployment. Ethnic minorities formed 46% and were bilingual or learning English.

Sample
The COGS was part of a Family Numeracy Project (FNP). 13 pupils were selected as parents participated in the FNP to improve skills and support learning.

How was COGS organised?
Foundation Stage 1 and 2 children (FS1 & 2) had the COGS for 30 minutes twice weekly. The Sage Assessments of Language and Thinking (SALT) (interview and story retelling) were used to establish narrative thinking and communication levels (Sage, 2000a). Goal 1 was pursued with both FS1 & 2 groups, but differentiated for need. The researcher and course leader team taught planning and facilitating sessions with a ratio of 2:5 (FS1) and 1:4 (FS2) adults to children.

Why was COGS relevant?

Targeting narrative competence was necessary as Ofsted identified many entering the FS with attainment well below norms in language, literacy and social development. Also, the FS1 group had many communication challenges, including 2 children learning English and 1 each of hyperlexia, dysfluency, dyslalia and mild learning difficulties. In FS2, 3 pupils were learning English and 5 lacked the confidence to express themselves. A summer-born Y1 child had not adjusted to a formal, FS2/Year 1 setting and wanted to play.

Methodology

An action research approach aimed to improve abilities. Planning, teaching and researching the COGS simultaneously enabled insights into the process. Plans were co-constructed between course leader and researcher, reflected on and evaluated after sessions.

Research design

The aim was to replicate previous COGS studies by collecting quantitative data in a quasi-scientific design, employing pre- and post-teaching measures with a control group and qualitative evidence from children, staff and parents. This proved impossible and a design, without controls, used tests before and after teaching. Data was strengthened by non-participative class observations of children after the project. Although policy considers large-scale research, small studies build evidence for richer, meaningful teaching and learning.

Group Results

Table 3 shows that narrative thinking increased, measured by SALT tests, before and after intervention. Scores vary across tests. In interviews, when *content and convention* scores combine and compare with *clarity* and *conduct,* greater progress is in the former. Increases in achievement on the post-retell test reflect a few not responding on pre-testing.

To what extent was progress in narrative thinking and structure due to the COGS?

A lack of controls means claims that positive results are solely due to the COGS are limited. Other factors are possible beneficial effects of: *participating in extra numeracy lessons; parent involvement in numeracy lessons; extra attention at home; and lesson engagement.* As children were observed in classes, discussion centres on why progress in narrative competence was more likely to have developed from COGS than usual learning contexts.

Table 3 – Group 1 & 2 SALT Scores Before and After Intervention

Test		Test mean totals	Raw score increase	% increase
Total scores SALT narrative retell	Before	6.625	8	120.7
	After	14.625		
Total scores SALT interview	Before	54	14.25	26.3
	After	68.25		
Content and convention scores only from SALT interview	Before	20	10.2	51
	After	30.2		
Clarity and conduct scores only from SALT interview	Before	34.66	3.44	10
	After	38.1		

Comparing COGS pedagogy with that observed in classes

Class observations noted key differences between the COGS approach and daily lessons. This might suggest that development of narrative competence was more likely in one context than the other. Learning principles and teaching focus in the COGS were extrapolated and located in classes. Children were observed in three settings over four days, which, although limited, is more than the time spent by Ofsted inspectors. 5 children were in FS1 and 4 in FS2, with the remaining 4 in a mixed FS2/ Year 1 class. Evidence collected from this protocol is in Table 4.

Compared with the COGS, other settings showed little evidence of providing children with opportunities for '*sustained shared thinking*', with a lack of focus on '*message-orientated communication*'. In FS1, structured play extended by adult intervention (*a hallmark of good practice*) was not observed, nor facilitation of group work. In FS2, and to a greater extent in the mixed FS2/Year 1 group, '*sustained shared thinking*' was not seen. It can be argued this was due to *formal* teaching, giving few opportunities for *informal* interactions to build confidence and competencies. Studies suggest that for many children the shift between informal and formal learning begins between FS1 & 2 (McInnes, 2002). A formal approach to literacy and numeracy appeases Year 1 colleagues (Adams *et al*. 2004, Sage, 2017). An early formal approach may have a negative impact on child involvement (McInnes, ibid). COGS are a '*halfway house*' between informality of home and school formal communication, providing chances to co-construct understanding of events and learn with others. It can be argued that this context was more instrumental in developing narrative thinking and communicative competence.

Table 4 – Comparing Learning Opportunities in COGS with Examples in Lessons

Aspect	Settings				
	COGS GRP	Year 1/FS2 Class	FS2 Class	FS1 Class	
Clarity Coached in voice techniques Speak and present in front of peers	Part of teaching objectives Embedded in activities for all children	Incidental Done by a few children in literacy	Incidental Done by a few children in literacy	Incidental Done by a number of children in circle time	
Content Verbalise and develop a number of ideas Think laterally and creatively	Embedded in activities for all children Embedded in activities for all children	Rarely observed in whole class situations. Done by one child in practical maths. More opportunities for small groups of children working with classroom assistant. Observed once in practical maths	Rarely observed in whole class situations. Done by whole class during an assembly. Observed during an assembly	Children's ideas elicited more during teacher's small group time, and when children could work individually with an adult. Observed in structured play	

(continued)

Table 4 – (continued)

Aspect	COGS GRP	Year 1/FS2 Class	FS2 Class	FS1 Class
		Settings		
Convention	Embedded in activities for all children	Children's own interests incidental	Children's own interests incidental	Structured play and circle time tended to be embedded in children's own interests.
Activities grounded in children's interests	Embedded in activities for all children	Part of the plenary encouraged children to listen to peers, but did not involve interaction.	Part of the plenary and assembly encouraged children to listen to peers, but did not involve interaction.	Structured play enabled children to interact with each other but due to the high number of children to adults these interactions were rarely structured.
Structured activities that encourage children to actively listen to and interact with peers	Some activities were structured to scaffold children's questioning	No questions, except procedural ones observed.	No questions, except procedural ones were observed.	No questions, except procedural ones were observed.
Ask questions of their peers and the teacher	Sensitively elicited by the teacher	Not observed	Not observed	Not observed
Evaluate their peers' performances and ideas				
Conduct	Embedded in the purpose of some of the activities	Incidental	Incidental	Incidental
Made aware of the importance of eye contact, facial expression and gesture	Embedded in the purpose of some of the activities	Not observed	Not observed	Not observed
Encouraged to make their message interesting to the listener	Questions open, all children to answer	Questions were closed and only some children answered during the whole class part of the lesson.	Question were closed and only some children answered during the whole class part of the lesson.	Some children answered their teacher during small group time.
Answer questions from their teacher	Activities scaffolded children to answer questions from their	Observed during informal choosing time	Observed during independent, small group work	Children interacted informally during play activities, but were not observed formulating or answering each other's questions
Answer questions from their peers				

Reflection

For narrative thinking and communicative competence to develop certain conditions must be created, evidenced in the COGS but not in classes. This indicates that teachers and pupils need more narrative knowledge and competence to enhance teaching and learning. It suggests a less formal, prescriptive approach to the KS1 curriculum, as this skews how teachers feel children should be taught in FS2 for Year 1. In Wales, the picture is different as a new FS, to extend play-based learning for 3-to-7-year-olds, was implemented from 2008 (QCAAW, 2004), showing how UK provision differs. Therefore, a key question is: how can the COGS shift from a useful intervention to embedment in lessons? This is feasible, because although the COGS took place in small groups, there is evidence of its success in whole-class teaching (Sage, 2017). It contributes to a better balance between child- and adult-initiated activities. The possibility is only likely if a gentle shift between informal and formal learning is achieved. Concerns about transition between FS2 and KS1 are voiced. Sanders *et al*. (2005) suggested that more play should be extended to those in Year 1. Although this is not as radical as Welsh approaches, it shows a swing against excessive formality.

REVIEW

The summary is written just as figures suggest that there is a 13% decline in students taking English language and literature as an advanced subject in Britain. This accompanied an article by Pip Sloan in the *Telegraph*: 'Is My English Degree Really Such a "Mickey Mouse" Subject?' (23 July 2019). It resulted in experts pointing out that all employers rate English and the competencies it teaches as vital for successful performances, as, of course, is the study of other languages which assists understanding of others and employability in plural societies. Google's Project Oxygen, launched in 2008, is a data-intensive study to understand the qualities that lead to life success. The top abilities identified were communicating and listening well, possessing insights into different values and points of view, showing empathy towards others and being a critical, creative problem-solver to connect complex ideas. These competencies are gained by studying English language and literature and that of other linguistic cultures. Technical abilities, like coding, may achieve a first job but will not progress a career, which depends on personal attributes. In the UK, we do not regard the teaching of English and communication or the acquisition of other languages as being as important as scientific subjects. This is not so in the top 10 nations in education league tables, who focus on students achieving narrative language and thinking in message-orientated communication. Employers bring limited communication to our attention and reports (UNESCO, 2013–16) indicate this is a reason for British performance being below comparable nations. Therefore, it is time we valued English as the medium of international exchanges, along with the learning of other languages to facilitate tolerance and understanding.

MAIN POINTS

1. English dominates globally through Commonwealth links and international language status.
2. The 7000+ world languages belong to families, with many varieties of form which make second-language acquisition difficult unless belonging to the same group.
3. The importance of language to thinking and action requires top status in teaching across all subjects and an approach built on its narrative development, like the COGS.
4. Decline in students taking advanced English qualifications is concerning, as employers and researchers assert its importance for learning the qualities leading to life success.

REFERENCES

Adams, S., Alexander, E., Drummond, M. & Moyles, J. (2004) *Inside the Foundation Stage: Recreating the Reception Year.* London: ATL

Alexander, R. (2000) *Culture and Pedagogy: International Comparisons in Primary Education.* Oxford: Blackwell

Alexander, R. (2003) *Talk in Teaching and Learning: International Perspectives. New Perspectives on Spoken English: Discussion Papers.* London: QCA

Anderson, S. & Lightfoot, D. (2002) *The Language Organ: Linguistics as Cognitive Physiology.* Cambridge: Cambridge University Press

Anderson, S. (2012). *Languages: A Very Short Introduction.* Oxford: Oxford University Press

Baker, M. (2001) *The Atoms of Language.* New York: Basic Books

Bell, N. (1991) *Visualising and Verbalizing*, Revised Edition. Paso Robles, CA: Academy of Reading Publications

Boyer, E. (1992) *Read to Learn: A Mandate for the Nation.* Princeton, NJ: The Carnegie Foundation for the Advancement of Teaching

Brigman, G., Lane, D. & Switzer, D. (1999) *Teaching Children School Success Skills. The Journal of Educational Research* 92 (6), pp. 323–329

Bruner, J. (1990) *Acts of Meaning*. Cambridge, MA: Harvard University Press

Butzkamm, W (2000) *Message Orientated Communication*. In: Michael Byram (Ed.) *Routledge Encyclopedia of Language Teaching and Learning*. London & New York: Routledge Educational Research.

Bybee. J., Perkins, R. & Pagliuca, W. (1994) *The Evolution of Grammar: Tense, Aspect and Modality in the Languages of the World*. Chicago: University of Chicago Press.

Chambers, J. & Trudgill, P. (1998) *Dialectology*. 2nd Edn. Cambridge: Cambridge University Press.

Frazier, I. (2001) *Narrative Modes of Thinking Applied to Piano Pedagogy*. *Piano Pedagogy Forum* 4, (2). keyboardpedagogy.org/images/PPF/PPF%20Vol.%204-5.pdf (accessed 29 July 2019)

Frid, I. Ohlen, J. Bergbom, I. (2000) *On the Use of Narratives in Nursing Research*. *Journal of Advanced Nursing*. 32 (3). pp. 695–703. www.ingentaconnect.com/content/bsc/jan/2000/00000032/00000003/art01530 (accessed. 29 July 2019)

Google's Project Oxygen (2018 onwards) *A Case Study in Connection Culture*. www.hrexchangenetwork.com/hr-talent-management/articles/google-a-case-study-in-connection-culture (accessed 25 July 2019).

Intermar (2014) *Communication & Employment Issues in European Military and Marine Academies*. Published by the EU Commission

Hargreaves, D. & Hargreaves, L. (1997) *Children's Development 3–7: The Learning Relationship in the Early Years*. In Kitson, N. Merry, R. (Eds.) *Teaching in the Primary School: A Learning Relationship*. London: Routledge

Hayward, G. & Fernandez, R. (2004) *Oxford Review of Education* 30 (1), pp. 117–145

Hogarth, T. Shury, J. Vivian, D. Wilson, R. (2001) Employer Skill Survey 2001. London: DFES

Lees, J., Smithies, G. & Chambers, C. (2001) Let's Talk: A Community-Based Language Promotion Project for Sure Start, in proceedings of the RCSLT National Conference: Sharing Communication, Birmingham, April, 2001

Mehrabian, A. (1971) *Silent Messages*. Belmont, CA: Wadsworth

McInnes, K. (2002) *What Are the Educational Experiences of 4-Year Olds! A Comparative Study of 4 Year-Olds in Nursery and Reception*. Early Years, 22 (2), pp. 119–127

Mills, C. & Mills, D. (1998) *Britain's Early Years*. London: Channel 4 Television. Cited by Sharp, C. (1998) *Age of Starting School and the Early Years Curriculum: A select annotated bibliography.* NFER. www.nfer.ac.uk

Qualifications, Curriculum and Assessment Authority for Wales (2004) *The Foundation Phase in Wales: A Draft Framework for Children's Learning.* www.accac.org.uk (accessed 29 July 2019)

Romaine, S. (2000) *Language in Society.* 2nd Edn. Oxford: Oxford University Press.

Sage, R. (1986) *A Question of Language Disorder.* MRC Trent Research Report.

Sage, R. (2000a). *Class Talk: Effective Classroom Communication.* Stafford: Network Educational Press

Sage, R. (2000b) *COGS: Communication Opportunity Group Scheme.* Leicester: University of Leicester

Sage, R. (2002) *"Successful Students". A project to teach communication skills and support learning.* Leicester: University of Leicester

Sage, R. (2004) *A World of Difference: Tackling Inclusion in Schools.* Stafford: Network Educational Press

Sage, R. (2017) (Ed.) *Paradoxes in Education.* Rotterdam: SENSE International Press

Sanders, D., White, G., Burge, B., Sharp, C., Eames, A., McEune, R. & Grayson, H. (2005). *A Study of the Transition from the Foundation Stage to Key Stage 1.* (DfES Research Report SSU/2005/FR/013). London: DfES.

Sutin, A., Costa, P., Miech, R. & Eaton, W. (2009) *Personality Career Success: Concurrent & Longitudinal Relations. European Journal of Personality,* 23 (2), pp. 71–84

Sylva, K., Melhuish, E., Sammons, P., Siraj-Blatchford, I. & Taggart, B. (2004). *Effective Provision of Pre-school Education (EPPE) project: Final Report.* London: DfES. www.dfes.gov.uk (accessed 29 July 2019)

UNESCO (2013–2016) *1. Education for All. 2. Children's Well-being in Rich Countries.* Paris: UNESCO Headquarters

Vygotsky, L.S. (1962) *Thought and Language.* Cambridge, MA: M.I.T. Press

Wang, M., Haertel, G. & Walberg, H. (1994) *What Helps Students Learn? Edu.Leadership,* 51 (4), pp.74–79

Weinreich, M. (1980) *History of the Yiddish Language.* Vol. 1. New Haven, CT & London: Yale University Press

Wood, D. (1998) *How Children Think and Learn.* Oxford: Blackwell

4

MULTIPLE LITERACIES AND PERSPECTIVES IN A GLOBAL SOCIETY

PREVIEW

'Multi-literacies' (new literacies) assume that individuals make sense of information by means other than traditional reading and writing. This includes linguistic, visual, audio, spatial and gestural ways of making meaning. People in modern societies need to learn how to construct knowledge from multiple sources and modes of representation (International Bureau of Education 2012).* This chapter discusses a broad based conception of literacy, including all verbal and non-verbal ways of symbolising and representing meanings, highlighting their importance in personal and academic success. Over the past century, the world has moved from spoken to written communication with mass education regarding reading and writing as the hallmark of achievement. Success in written language depends, however, on literate structures achieved in formal speaking, which assemble quantities of information for instructing, informing and explaining. Today, we spend more time looking at visual display screens than talking formally, with strong evidence that a lack of speaking opportunities affects verbal comprehension and expression, and so all forms of learning. Human development indicates that formal talk structures must be achieved to shift primary language

---

*Definition of Multi-literacies: IBE-UNESCO: International Bureau of Education. www.ibe.unesco.org/en/glossary-curriculum-terminologym/multiple-literacies (accessed 3 August 2019) *(spoken) into secondary modes (written). Many people show difficulties in both spoken and written communication, suggesting more awareness of these processes in teaching (GRA Survey, 2009). Looking at a range of interdisciplinary perspectives is the best way to facilitate learning.*

THE CONTEXT

Today, the movement of people around the world is unprecedented. In the book *Meeting the Needs of Students with Diverse Backgrounds,* a visit to a local supermarket is mentioned, where 15 different languages were heard spoken before counting stopped (Sage, 2010). Such daily events demand awareness of others and an enhanced ability to communicate across a myriad of differences for better understanding of individual values and needs. It is essential to consider the nature of communication and develop subject knowledge in line with current needs for greater awareness and learning competence.

WHAT IS LITERACY?

Over the past century we have seen the move from spoken to written cultures and a consequent drive for mass education and universal literacy (*reading and writing competence*). In human development there has been a transition from non-verbal to verbal communication and then literate forms, basic to formal education. Multi-literacy combines competence to read, write and search for information, involving oral, digital, media and global abilities. Communication tools, emerging technologies, social and cultural forces constantly redefine what we mean by '*literacy*'. Students now need to use a broad range of literacies to achieve immediate learning objectives as well as recognise and develop creative possibilities. Many are collaborative writers and content creators in the digital world, providing chances to develop multiple literacies. In doing so, they reach a deeper understanding of the global community. Literacy types are continually expanding: *traditional, information, media, visual, cultural, digital and critical*, to name just a few!

Today, '*literacy*' refers to more than the ability to interpret printed text (e.g. Googling). The term '*literacy*' refers to emerging global modes of communication technologies that can be used for educational purposes. These include, but are not limited to, video, motion picture production, television, audio broadcasting, live video conferencing and multiplayer educational gaming – providing many ways that individuals can access and share information. Speaking, with its non-verbal systems signalling meaning, has less prominence in learning. Whitehead (1997, p. 177–78), however, points out literacies suffer '*if not established on a broad and deep foundation of worthwhile experiences of symbolising and representing meanings through non-verbal communication'.*

Marking meaning, non-verbally, are *voice dynamics, facial expressions, gestures, movements, postures, positioning, appearance and manner* – influenced by attitudes, values, intellect and personality. These attributes are seen in *listening, talk, dance, music, art, storytelling, poetry, drama, sport, scientific and mathematical investigations, rituals and religious celebrations*. Writing is one mode with many *'literacies'* and symbolic languages to be experienced, established and understood before becoming competent communicators.

MULTI-DISCIPLINE PERSPECTIVES IN LEARNING

We need to look at learning holistically and consider contributions that various disciplines can make in an age when many people are not educated in their mother tongue.

Neuropsychology and second language acquisition (SLA) discuss three major issues:

1. Localisation of Language 1 (L1) and/or L2 in the Left (L) and Right (R) brains
2. Representation of different characteristics (*e.g. English with phonetically based scripts versus languages with ideographic representations, as in Chinese and Japanese*)
3. A critical period hypothesis for L2 acquisition

Current views suggest that the L brain processes analytically and serially and the R brain holistically and spatially, although this may not be absolute. Genesee (1988) discussed three factors: *age, level of proficiency and manner* of SLA. The *age* hypothesis states there will be more R brain involvement in SLA the later it is learnt. The *stage* hypothesis has little empirical backing, as R brain involvement in SLA is more evident in less proficient bilinguals. The *manner* hypothesis suggests greater R brain involvement in languages learnt informally, or conversely the case, with the former situation mostly supported by experts. This suggests learning through real experiences, motivated by personal need.

The issue of language-specific effects in bilinguals resulting from 1) phonetic or ideographic script, 2) L-R or R-L script direction and 3) appositional or propositional language mode, suggests the following:

- ***propositional languages,*** using analytical, abstract thought, like English, may be associated with L brain processing, as are phonetically based scripts. *In contrast…*

- *appositional languages,* using holistic, integrated thinking, like Navajo and Hopi, may be associated with R brain processing, as may be the case for ideographic scripts.

There is little empirical research on script direction and differential brain processing. The *critical period hypothesis* (CPH), suggesting a stage in development when SLA occurs easily, is supported as pre-pubertal learners achieve higher, more native-like L2 proficiency, although this is not equivocal. Such knowledge informs us about the mechanics of SLA to reflect on teaching approaches. Although we may be a long way from fathoming brain complexities, neuroscience is clarifying roles of the L and R brain in learning, indicating that languages depend on different lead sides for their performance.

PSYCHOLINGUISTIC ISSUES IN SLA

Chomsky's (1957) *transformational-generative grammar* theory led to abandoning the behaviourist view of language as a set of habits, considering that it is represented by abstract rules internalised by speakers from what is heard in context. This paradigm is difficult to reconcile across languages. For example, the native Australian language, Warlpiri, has grammar elements everywhere in a sentence and is not neatly packaged, as in '*universal grammar*'. If you look at Urdu, now spoken in the UK, the sentence subject is used differently to the way it is treated in English. Spanish forms sentences without separate subjects, as in *Tengo* (*I have*), in which the person 'I' is indicated by 'o' at the end of the verb. The Amazon language Piraha does not use recursion (*phrases tucked inside others*) to build sentences (Sage, 2018).

As a speech therapist assessing communication, there is a great deal more to consider than just innate grammar. Speaking abilities to categorise (*people/objects*) and understand relations among things are fundamental. One must also grasp the intention of others, as well as memorise and integrate words, voice, gestures, personal characteristics, manner and context clues to enable successful language. This is accompanied by social abilities to connect with others (*phatics*) and develop trust, coping with different views, values and attitudes positively. Since words convey only 7% of affective message meaning, voice dynamics (38%) and gestures (55%) are crucial elements. These latter paralinguistic elements build a foundation for effective verbal language understanding and expression (Mehrabian, 1971, Sage, 2000). In development, communication comes first and grammar later, when thinking becomes *'externalised'* for conversation. Language cannot exist without specific human forms of social cognition and interaction.

Corder (1967) has further directed us to consider a learner's systematic errors and ignore random mistakes. Krashen (1976, 1982) demonstrates that most errors are not due to L1 influence, as traditionally thought. Seliger (1983) focused on how the learner develops language and the processes involved, introducing *transitional competence (changing state of flux as new knowledge is acquired)*, *approximate* systems *(progressive learning)* and *interlanguage (unique grammar)*. Language and training transfer and overgeneralisation are viewed as hypothesis testing.

Affective factors lead us to consider psychological characteristics of the successful learner – *someone with drive to communicate, willing to take risks, appear foolish and able to live with vagueness.* Schumann (1975) makes *affective* factors the central focus of his *acculturation* model, as the initiator and controller of the process. Krashen's *monitor* theory (1982) deals with paradoxes that exist in informal or formal learning, as when abilities vary according to conditions (*e.g. writing better than speaking*). He makes distinction between learning (*formal*) and acquisition (*natural*). Learned grammar can be accessed only when the rule is known with time to think about form, using the monitor to edit output. Krashen's *input hypothesis* suggests we develop by being exposed to input beyond that known. Criticism centres on whether the model lacks factors known to affect SLA, like a speaker's desire to communicate, so that grammatical factors assume a minor role. Psycholinguistics provides no exact answers but does lead us to question and accept the immense complexity in language learning, as well as acknowledging the differing performances of individuals according to conditions.

SOCIOLINGUISTIC AND SOCIAL PSYCHOLOGICAL APPROACHES TO SLA

There has been attempt, in research, to answer the following questions:

1. What causes interlanguage variation?
2. Does interlanguage change over time?
3. What is the role of sociolinguistic transfer in SLA?
4. What is the sociolinguistic role in L2 development?
5. What is *communicative competence*?

Beebe (1988) says Labov's *attention to speech* paradigm (1981) does not adequately explain *style shifting* in speaking, and describes the *gradual diffusion model* (Gatbonton-Segalowitz, 1975) in which SLA has two phases: 1) when a new rule is used simultaneously with an incorrect one; 2) where the correct rule begins to *'win out'*. The notion of *communicative competence* – knowledge of how to use language appropriately in context (Hymes, 1972) – is based on speech acts (*e.g. compliments, requests, apologies, etc.*) and how verbal and non-verbal languages fulfil social functions. Giles (1980) and others went on to elaborate *speech accommodation* theory to explain talk variation and adjustment, as when a speaker makes utterances more similar to their interlocutor (*convergence*). *Speech maintenance* is failure or refusal to converge, called *divergence*.

Speech accommodation involves shifts in all linguistic domains, such as phonology, syntax, grammar, lexicon, content, speech rate, etc., and is often the same as Labov's *style shifts*. *Theories of similarity attraction, social exchange, causal attribution* and *intergroup distinctiveness* are combined in *social accommodation,* to explain the social psychological processes involved. Also, *attitudes* and *motivations* explain differential success (Lambert, 1979). Subjects evaluated bilingual voices in both languages to determine attitudes towards native speakers of French and English in French-speaking Canada. Initial studies indicated an *integrative orientation (desire to integrate with the target culture)* but later ones showed *instrumental orientation (need to know language for work)* as effective. Both *instrumental* and *integrative* orientations are viewed as necessary for successful learning in general.

There is much applicable research in sociolinguistics and social psychology, and this brief summary by no means does it justice. Reactions to speech, particularly dialect sensitivity and pragmatic ability, like deference and politeness, are important to pursue when considering communication and language acquisition. This is vital for teachers who speak and act in a different way to the students they teach. Many students have problems with a communication style that is unfamiliar for both mental and emotional reasons. We naturally fear others who look, speak and behave differently to us.

Pedagogical issues and SLA

Evidence that *instructed interlanguage* and *natural development* are similar surfaced with Krashen *et al.* (1976) claiming morpheme usage was similar regardless of first language background. He prescribed *comprehensible input* as necessary provision for SLA to occur. Pienemann (1984) found that instruction made learners progress faster if they were ready for it, but had no effect if they were not, so acquisition sequences may well be immutable. This is important in education, as prescriptive regimes mean students are often forced to learn things that they are not able to acquire at the designated time.

Long (1985) cites evidence to suggest the effect of instruction on levels of attainment, as focus on form makes it likely that items acquired were attended to by learners. Supporting this, Schmidt and Frota's diary study (1986) noticed forms outside class after being taught. The conclusion is that instruction benefits the processes, rate and ultimate level of attainment, but the sequence in which forms are acquired remains unaltered.

Cummins (1988) introduced the bilingual policy debate, originating in the 1960s, when proponents argued that such approaches were necessary to promote a healthy self-concept. He clarified the different meanings attached to the word *immersion*, with programmes often demonstrating *submersion* and exclusive use of L2. He gives a useful guide to characteristics in order to make judgements and comparisons, whether: 1) L1 or L2 is first emphasised; 2) teacher and programme are bilingual or monolingual; and 3) participants are majority or minority students. Research refutes a claim that minority students suffer academically if required to study in their weaker language.

TWO THEORIES EXPLAIN WHAT WORKS IN BILINGUAL PROGRAMMES

1) *Cognitive academic abilities* developed in L1 transfer to L2 because they involve a *common underlying proficiency*. Transfer from the minority to the majority language is more likely because exposure and motivation are more favourable than the reverse.

2) *Sufficient, comprehensible input generalisation* asserts that interesting and relevant, rather than grammatically sequenced input is necessary for SLA (Krashen, 1982). It focuses on the *'here and now'* with repetitions, confirmation/comprehension checks and clarification requests. Bilingual or immersion programmes are more successful than traditional ones due to comprehensible, relevant input with L2 as the mode of instruction. They provide modified L2 input and promote L1 literacy which transfers to L2.

COMMENT

Language, like any subject, benefits from holistic approaches that employ information culled from relevant disciplines. We must build all education on knowledge gained from experience, and, just as the learner needs to practise the target goal, so the teacher must study the academic disciplines that inform pedagogy. Neuropsychology makes us aware of the utility of employing both L and R brains in learning, focusing on the creative nature of communication as well as its social and linguistic rules. From psycholinguistics we note the importance of teacher awareness of sentences with similar surface structure but different deep structure, such as:

1. *He asked me to be his teacher.*
2. *He considered me to be his teacher.*

According to Chomsky's *transformational-generative grammar* (extended standard theory), a teacher could be deluded by similar surface structure similarities. The verbs, however, are merely *homonyms*, with different deep structures evident as below:

3. *He asked me to become his teacher.*
4. *He considered me to become his teacher.*
5. *He considered me his teacher.*
6. *He asked me his teacher.*

Two different verbs are evident in sentence 1 and 2, with 1 having the same meaning as 3 with *to be* and *to become* equivalent. It is not possible to substitute *to become* for *to be* in 4, because it is a different structure. *To be* in sentence 2 can be deleted as a complement in 5, but in 1 this would be ungrammatical, as in 6. Also, errors are mostly intelligent attempts to master language and should not be criticised but paraphrased correctly. In so doing we deal with learner difficulties by showing them a range of ways to communicate the same idea (Sage, 2018).

Regarding the sociolinguistic perspective, there are several teaching implications, like the systematic, regular, rule-governed nature of interlanguage, even when containing errors. Labov's paradigm of high attention to speech form, however, may produce low accuracy, leading to hesitancy or errors in natural communication. High attention to form is most appropriate for learners with low linguistic knowledge. Pedagogical perspectives are rich in implications, indicating that formal instruction makes a difference and focus on form produces accuracy. Pienemann's (1984) *learnability-teachability hypotheses* suggest that teachers should carefully monitor the gap between what they teach and what students learn. There is value in stepping back, reflecting and facilitating group work, which encourages students to learn from one another and teachers to observe. Bilingual educational perspectives advise us not to cling to conventional wisdoms that are not well founded and put evidence into a coherent theory of practice.

There must be a critical re-examination of traditional ideas with a creative implementation of new ones. We should integrate experimental knowledge and experiential information and see things in a theoretical context. Unity of view lies in a diversity of knowledge and ability to assemble and implement it holistically. This is demonstrated in a unique British immersion programme, *The Glain Orme Language and Business Course*, designed for adults requiring English as their working language

(Wilson-Draper, 2019). It is a novel approach to teaching, in the experience of participants, who mainly come from European countries and have been used to a traditional emphasis on syntax and grammar, whereas the goal must be to teach speaking and writing that is appropriate for the context. Learning is developed from real experiences such as shopping, cooking, hosting guests, etc. Studies have shown that intelligence level is a good predictor of French and English academic proficiency (Genesee, 1984), but not of conversational ability, shown in the performance of the participants on this programme, if new to the methods.

Until the 1950s, language acquisition was largely influenced by behaviourism, when Chomsky's work marked a turning point regarding its nature and acquisition. The claim that language is learned by acquiring habits was challenged by Chomsky's assertion that humans produce novel utterances when speaking rather than imitating what they hear. Chomsky claims we possess an innate language acquisition device (LAD) to abstract rules from the incomplete utterances of normal speakers. Others, like Vygotsky, suggest that language depends on the social context, with interaction playing the major role in acquisition. The notion of *communicative competence* reinforces this, by embracing a range of verbal and non-verbal abilities appropriate for the context in which events happen. This targets a broader perspective than just grammatical knowledge, which can only be realised from real and relevant social experience based on need (Sage, 2018).

There has been a conscious effort to replace traditional transmission teaching with more interactive approaches, but these are hampered by prescriptive governmental requirements. In our global world we engage with many cultures and languages and a wider range of printed and televisual texts outside educational establishments than within. It is essential that educators acknowledge such influences, recognising and responding to this multi-stranded experience with flexible approaches that meet needs of individual learners rather than educational systems.

LEARNING FOR THE NEW WORLD

There is more than one way to... '*skin a cat*'? Yes, there is also more than one way to transmit and receive information. For students, there are many methods to search for and obtain information they need. Likewise, there are numerous ways they can communicate knowledge. In most schools, the Media Centre is the primary information hub. Strategies for incorporating multiple literacies into practice are now discussed:

Multiple literacies are abilities to interpret the many formats, sources, or media through which we obtain information. The four competencies comprising these are *visual, textual, digital* and *technological*. **Visual literacy** is how we understand and interpret visual representations of knowledge. **Textual literacy** refers to how we understand and interpret text and communicate through written means. **Digital literacy** is the ability to locate, evaluate and synthesise digital sources (*online*). **Technological literacy** is using technology appropriately and responsibly. The digital age has transformed ways we seek information. Gone are days of thumbing through paper to locate library sources. Teachers now must expose students to information through multiple literacies, while teaching **digital citizenship** – responsibilities for those using technology to ensure appropriate use.

INTEGRATING MULTIPLE LITERACIES

How can we incorporate strategies and practices to enhance student abilities in multiple literacies? Below are practices to accomplish this.

Use physical space to convey information in various formats.
Physical space provides many possibilities. Students need help to locate, read and understand information from many sources. You can use walls and a physical arrangement to support this goal. There may be wall posters encouraging students to read. While these have value, replace some with infographics (*see image below*) that communicate information students should know, like abilities needed for today's world. You could also ask them to *create* their own infographics on a topic and turn this into a contest, or find other ways to focus attention. Also, consider incorporating *information nooks*. You may have *reading nooks* for students to relax and read a book. What about spaces that cater specifically for those who like to listen to podcasts, read an ebook or listen to an audiobook? These media are all forms of information and strengthen multiple literacies whilst catering for varying needs. In Japan, there are many mantras on the walls in corridors – *Work Hard, Play Hard, Think of Others, not Just Yourself,* etc. Classroom walls are bare, so as not to distract, but corridors are often bright and informative.

An Infographic Example

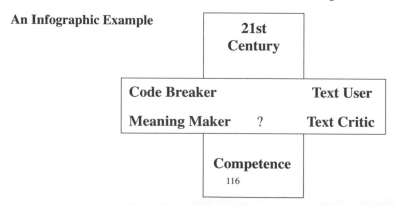

21st
Century

| Code Breaker | | Text User |
| Meaning Maker | ? | Text Critic |

Competence

What does research tell us about teaching, learning and multiple literacies?

1. Supported engagement with multiple literacies increases motivation and success

Covering daily learning activities for school goals, Hinchman and Sheridan-Thomas (2008) review key topics in literacy instruction and provide research-based recommendations for practice by leading scholars. These are culturally responsive – using multiple texts and digital media; integrating literacy instruction with science, social studies and mathematics; teaching English language to learners and struggling readers alongside other subjects. Case studies, discussion, questions and activities in each chapter, with detailed ideas for lesson planning, make this a useful resource and professional development tool. Students are motivated through engagement with multiple literacies of their choice. Best practices must account for '*the impact of texts of all kinds (visual, print, digital, sound, multimodal, performance) on young people's identity-making practices and, especially, on how text mediates young people's perceptions of themselves as literate beings*' (p. 15).

The Pew Internet Study (Lenhart *et al.*, 2008) says that students do not think of emails and text messages as writing. This disconnect matters because they believe writing is essential for success and that more school instruction would help them. The survey found:

1. Even though learners are heavily embedded in a tech-rich world, they do not believe that communication over the Internet or text messaging is writing.
2. The impact of technology on writing is vital because most believe that this is important to future success and enhanced by sharing this orally.
3. Students are motivated to write on relevant topics, achieve high expectations, have an interested audience with opportunities to write creatively.
4. Writing in school is a daily activity, but most assignments are short.
5. Students believe that the writing instruction they receive in school could be improved, and non-school writing is widespread amongst them.
6. Multi-channel students and gadget owners do not write any more or less than their counterparts, but bloggers are more prolific.
7. Teens more often write by hand for both out-of-school writing and schoolwork.

8. Although tech-savvy, students do not believe that writing on computers make a big difference to quality.
9. Parents are generally more positive than their children about the effect of computers and text-based communication tools on writing.
10. Students enjoy non-school writing and to a lesser extent the writing they do in class.

Research argues that reading materials should tap students' diverse interests and a range of difficulty. Learners benefit from sustained engagement with a wide range of texts.

2. Multiple literacy practices are embedded in social understandings and students may need help seeing themselves as readers and writers. Social networks surround ways learners read texts. The embedded nature of these within their networks enable them to accrue social capital. Findings suggest that how to motivate students with active out-of-school literacies to engage with school mastery is unclear (Moje *et al.*, 2008). The study looks at what teens and parents report about technology and in- and out-of-school writing. Teens do not always perceive their out-of-school writing (*technology-mediated writing*) as writing. Struggling readers need instruction linking personal experiences with their texts, making connections between their existing literacy resources and the ones necessary for various disciplines.

3. Student choice and participation increases motivation
Research suggests that students require intrinsic motivation for learning. Motivation is encouraged from real-world connections enabling meaningful choices. A responsive classroom culture is key and encourages students to take ownership for learning. Teachers who understand and respond to student lack of motivation can help them take an active role in learning and maintain engagement. In effective schools, classroom conversations about what, how and why we read are vital curriculum inputs. Discussion-based approaches to content are strongly linked to student achievement. However, high-stakes testing, like school exams, is not only narrowing curriculum content, but also constraining instructional approaches. Limited 'one right answer' or 'main idea' models run counter to research findings, which call for a richer, more engaged approach to instruction that does not target right or wrong responses (Biancarosa & Snow, 2004).

The 21st century is unlike any other in history, and new technologies and multimedia are more dramatic than the transition from oracy to literacy. Think how our media culture, globalisation, increasing diversity, technologies, social media, digitisation, cyber life and huge leaps in medical technologies have radically changed lives. Also, the world human population has more than doubled since 1970. Such changes have created problems, as well as opportunities, which today's students will inherit. They face changes and situations which we cannot imagine. What does the unknowable future mean for education today? Fifty years ago, we were certain about how to teach students for their future success. This is no longer the case, as that education paradigm is now obsolete. What is the answer? We need a new learning model, and must prepare students for the present and future by teaching them 21st century competencies, expanding the literacy concept. The three Rs, reading, writing and arithmetic, are important but no longer sufficient! Today, 'literacy' has new meaning – the ability to communicate effectively in any form.

Communicating well is not just adeptness at basic processes like reading, spelling and drawing, but also means mastering higher-order linguistic competencies like interpretation, analysis, critical appraisal, evaluation and creation. When discussing literacy, we use *'texts'* in a new way. These used to be *textbooks*. **Today, a *text* can be any form of communication, including articles, novels, non-fiction books, movies, plays, posters, photographs, paintings, advertisements and even computer games – any medium used to communicate.** This change in literacy meaning came about from the explosion of new ways to transmit messages, as technology developed.

MULTIPLE LITERACIES CAN BE VIEWED IN TWO WAYS

1. **Modal Literacies:**
   Meaning expressed in linguistic, visual, audio, spatial and gestural representations.
2. **Literacy by Medium:**
   Focus is on forms like *'language'*, *'visual'*, *'computer'*, *'film'*, *'media literacy'*, etc.

To show this in practice, the first three are compared at basic and advanced levels.

Basic level of communication literacy

1. **Language Literacy:**
   Involves reading and writing words and comprehension of sentences.

2. **Computer Literacy:**
   Includes turning the device on/off and how to: use input/output features (*keyboard, printer, touchscreen, camera and mouse*); understand icons and menus; navigate the machine; find a file; explore the internet; start programmes and apps; etc.
3. **Visual Literacy:**
   At a basic level the viewer/artist is aware of various production techniques, such as line, shape, colour, space, texture, plus design principles like balance, emphasis, harmony, movement, perspective and contrast.

Compare and contrast each of these aspects in these two pictures:

1. Bay Sage on the sofa

2. Sonny Townsend is a 10-year-old boy at a London State school who has given permission for use of one of his pictures to illustrate abstract thinking

Advanced communication literacy

1. **Language Literacy:**
    Included are: higher levels of comprehension like understanding symbolism and irony; determining the purpose of an author and the reader's response.
2. **Computer Literacy:**
    Included is: expert knowledge of software like word processors, spreadsheets, databases, browsers, email and graphic programmes, device setup, maintenance and backup, extending to computer programming skills.
3. **Visual Literacy:**
    First there is analysis of the image. In picture 1 above, the head of the dog is central to the image. The focal point is the dog's eye, so we begin to question what she is looking at. Secondly, we look for symbolism and ideas that are being communicated. We could speculate that the owner is a dog lover who is trying to make us appreciate loneliness and expectation in pets, or a broader empathy for anyone who would like to see us more. Thirdly, we think about the meaning that the ideas have for ourselves and how we are changed by this. **How does the picture of a dog make you feel and what you do about this is an individual experience?**

    In picture 2, there are many images. Which one do you focus on first? Why? Is there a connection between the various very different components which make up the picture? Do they give you some understanding of the artist? Why are some representations coloured and others not? What is the overall meaning? What ideas are being communicated? What are your feelings when viewing the picture? What sort of impression has the picture made on you? What does the picture tell you about this talented 10-year-old? **How are you changed by viewing this piece of art? Has it made you think in a different way about abstract representations?**

Developing communication levels

Many literacy forms are taught at school because jobs require effective communication. As technology develops, people need even higher levels of communication, with the ability to cooperate and collaborate equally. A grounding in the various basic skills is crucial for developing higher-order thinking and literacy. These abilities (*analysis, interpretation and response*) are similar in different media, so we can use what we know in one to help us think about the others.

Basic skills are not sufficient

As technology develops and displaces people from present jobs, those not having both basic and advanced literacy will find it harder to get work. Not only that, they will find it difficult to retrain for jobs requiring multiple literacies.

TYPES OF LITERACY

The four primary areas – *visual, textual, digital and technological literacy* **– are expanded:**

Visual

Visual literacy is the ability to understand and evaluate information presented through images like pictures, photographs, symbols and videos. It means going beyond simply looking at the image and involves assessing the message it is trying to convey, or the feelings it is designed to evoke. Developing strong visual literacy means teaching students to observe and analyse images. They must be trained to perceive the image as a whole, for overall meaning, before noting details of what they see. Then they should think about its purpose. Is it meant to inform, entertain or persuade? Finally, students must be able to infer the image's significance. Visual literacy also includes the ability to express effectively through digital media. Not all students will become artists, but a practical application is putting together a visual presentation that accurately and effectively communicates ideas.

Textual

Textual literacy is what most people would associate with its traditional definition. At a basic level it is the ability to assimilate written information, such as literature and documents, and communicate effectively in writing. However, textual literacy goes beyond merely reading information. Students must be able to analyse, interpret, evaluate and reflect on what they have read. Textual literacy includes the ability to put what is read into context, review it and challenge if necessary. Analysing and responding to books, blogs, news articles or websites, through reports, debates or persuasive writing, builds textual literacy.

Digital

Digital literacy refers to an ability to locate, evaluate and interpret information found through digital sources, like websites, smartphones and video games. Students must learn to evaluate digital media critically, determine if the source is credible and identify the author's viewpoint and intent. Students can learn to recognise satire from samples on spoof websites like *The Onion* or *Save the Pacific Northwest Tree Octopus*. Older students benefit from reading various opinions on situations and examining news articles to determine which ones contain the least bias.

Technological

Technological literacy is the ability to use a variety of technologies (*social media, online video sites and text messages*) appropriately, responsibly and ethically. A technologically literate student understands not only how to navigate digital devices, but also how to do so safely, while protecting their privacy and that of others, obeying copyright laws and respecting cultural diversity, beliefs and opinions in society. To develop technological literacy, students need projects that require online research.

USING MULTIPLE LITERACIES IN CLASSROOMS

Teaching multiple literacies requires teachers to understand technology themselves. They must engage with colleagues in the technology that students use, like social media, blogging and gaming. Teachers must provide opportunities for students to develop multiple literacies to locate, evaluate and process information, communicating what they have learned to others regularly. Try these ideas for integrating multiple literacies in class:

Create Engaging Classroom Activities

Engage in activities to promote visual literacy, like *Five Card Flickr*. This website fosters creativity by making stories using photos. Provide students with six random photos or images. Ask them to write a word associated with each image, name a song that reminds them of them and describe what all have in common. Invite students to compare answers with peers and discuss findings.

Diversify Text Media

Provide many ways for students to interact with text, such as print books, audio and electronic formats. Allow students to listen to an audiobook while following a print version. Post infographics where students can read them and allow time to listen to podcasts and then discuss content with peers.

Access Digital Media

Ensure that students access a variety of digital media for collecting and creating information. They may wish to read blogs or websites, watch YouTube videos or use streaming services to research topics. They can create a blog, video or other digital media presentation to relay what they have learned for feedback from others. Prepare students for senior school and beyond by allowing them to choose a topic to research for the term or year. Guide students in learning to read web pages – identify the author, determine the credibility of the information and cite sources. Students can then use digital media (*or a combination of digital and print*) to create a presentation.

Use Social Media

If students are 14 and older, consider a classroom Twitter account or Facebook group. Use it to communicate with students and model safe, responsible, ethical use.

Multiple literacies resources

Apart from classroom integration, there are many resources for students to develop multiple literacies. They will naturally use many of these, such as gaming, the Internet and social media outlets. Libraries recognise multiple literacies, offering student assistance, like a free computer and Internet access, ebooks, audiobooks, tablet access and digital media workshops. Students can use free tools available on smartphones, digital devices or computers to explore multiple literacies. Suggestions are:

- iMovie for video creation
- GarageBand for creating podcasts, music or sound effects

REVIEW

Today I am looking after a dog for a neighbour, because family problems mean he cannot be given attention. When Rick rang to make the arrangement, I said I needed to go and pay the paper bill before 8 a.m., but '*ring when you are ready*' after this time. I meant '*ring my mobile*', but Rick thought it was '*ring my doorbell*', so turned up at 8.15 a.m. Fortunately, I was back from the newsagents! Rick had filled in the *information gap* differently to my intention. This is the great problem with words and their inevitable *information* and *opinion gaps*. In an age when numerous languages are spoken by our contacts, there are likely to be many misunderstandings.

Teachers have an enormous challenge today, as they probably have students who do not speak English as their first language in classes. This means they need flexible approaches and the employment of a range of technologies to meet diverse needs. Understanding what is available to support learning, amongst the many literacies now in operation, is an important requirement. Media specialists in schools have a vital role in teaching colleagues and students what can assist learning.

What we gain from looking at how people acquire subsequent languages is important for all subject teaching, as learning is a communicative process. Firstly, education must follow developmental norms to be effective. When moving from speech therapy into teaching, it was surprising that curricula did not always follow such sequences. Nurseries taught colour before shape, which was against what I had been told. Norms suggest that concrete (*shape*) should be introduced before abstract (*colour*) ideas. None of us have the monopoly of the truth, but I have found in high-achieving countries, like Japan and Cuba, they teach *shape* well before *colour* in curricula schemes. Secondly, students need structure and opportunities to reflect and review for secure learning. Thirdly, they must experience learning applied practically, involving all sense modalities, to understand fully and commit information to memory. Fourthly, the content must be relevant for interests and needs. From neuropsychology we understand the role of both the L and R brains in learning different things, with techniques, like the PACE (see Sage, 2000), encouraging hemispheric coordination. Psychology and psycholinguistics define learning stages and the role of language in thinking. Sociology demonstrates the importance of context, network and affective influences to understand how action varies. Advanced communication literacy, in all its forms, is, therefore, needed to survive our complex world that is difficult to master successfully and dependent on basic competencies being in place. There is much evidence to suggest this is not the case, and employing a greater range of communicative processes to provide appropriate learning that suits the interests, ability and pace of each student makes sense. This is something which teachers recognise but the system of education does not always support.

MAIN POINTS

1. Today we have a broad-based conception of literacy, including all verbal and non-verbal ways of symbolising and representing meanings, highlighting their importance in personal and academic success.

2. There are 4 main types of literacy – *visual, textual, digital* and *technological*.
3. Literacy requires both basic and advanced competencies for our complex world.
4. Perspectives from many disciplines are used for a holistic approach to learning multi-literacies, giving students more flexibility in learning.

REFERENCES

Beebe, L.M. (1988) *Five Sociolinguistic Approaches to Second Language Acquisition*. In L.M. Beebe (Ed.) *Issues in Second Language Acquisition*. New York: Newbury House

Biancarosa, G., & Snow, C. (2004). *Reading Next: A Vision for Action and Research in Middle and High School Literacy*. Report to Carnegie Corporation of New York. Washington, DC: Alliance for Excellent Education.

Bruner, JS (1966) *Toward a Theory of Instruction*. New York: Newton

Chomsky, N. (1957) *Syntactic Structures*. The Hague: Mouton

Corder, S.P. (1967) *The Significance of Learners' Errors. International Review of Applied Linguistics*, 5 (4), pp. 161–170

Cummins, J. (1988) *Second Language Acquisition within Bilingual Educational Programs*. In L.M. Beebe (Ed.) *Issues in Second Language Acquisition*. New York: Newbury House

Gatbonton-Segalowitz (1975) *Systematic Variations in Second Language Speech*. PhD Dissertation, McGill University

Genesee, E. (1984) *Historical and Theoretical Foundations of Immersion Education*. In California State Department of Education (Ed.) *Studies on Immersion Education:A Collection for US Educators*. Sacramento, CA: California Department of Education

Genesee, F. (1988) *Neuropsychology and Second Language Acquisition* in L.M. Beebe (Ed.) *Issues in Second Language Acquisition*. New York: Newbury House

Giles, H. (1980) *Accommodation Theory: Some New Directions*. In M.W.S. de Silva (Ed.) *Aspects of Linguistic Behaviour*. York: University of York

Graduate Recruiters Association (2009) *Employers Survey*. London: GRA

Hinchman, K & Sheridan-Thomas, H. (Eds.) (2008). *Best Practices in Adolescent Literacy Instruction*. New York: Guilford Press

Hymes, D. (1972) *On Communicative Competence*. In J.B. Pride and J. Holmes (Eds.) *Sociolinguistics*. Harmondsworth: Penguin Books

Krashen, S. (1982) *Principles and Practice in Second Language Acquisition*. London: Pergamon Press

Krashen, S, Sferlazza, V., Feldman, L. & Fathman, A. (1976) *Adult Performance on the SLOPE Test: More Evidence for a Natural Sequence in Adult Second Language Acquisition. Language Learning* 26, pp. 145–151

Labov, W. (1981) *Field Methods of the Project on Linguistic Change and Variation. Sociolinguistic Work Paper*, 81, pp. 1–37

Lambert, W.E. (1979) *Language as a Factor in Intergroup Relations.* In H. Giles & R. St. Clair (Eds.) *Language and Social Psychology.* Baltimore: University Park Press

Lenhart, A., Arafeh, S., Smith, A., Macgill, A. (2008). *Writing, Technology & Teens.* Washington, DC: Pew Research Center

Long, M.H. (1985) *A Role for Instruction in Second Language Acquisition. Task-based Language Teaching.* In K. Hyltenstam & M. Pienemann (Eds.) *Modelling & Assessing Second Language Development.* Clevedon: Multilingual Matters

Mehrabian, A. (1971) *Silent Messages.* Belmont, CA: Wadsworth

Moje, E., Overby, M., Tysvaer, N., & Morris, K. (2008). *The Complex World of Adolescent Literacy: Myths, Motivations and Mysteries. Harvard Educational Review*, 78 (1), pp. 107–154.

Pienemann, M. *Psychological Constraints on the Teachability of Languages.* Manuscript: U. of Sydney Dept. of German

Sage, R. (2000a) *The Communication Opportunity Group Scheme.* Leicester: University of Leicester

Sage, R. (2000b) *Class Talk: Successful Learning through Effective Communication.* London: Network Continuum

Sage, R. (2003) *Lend Us Your Ears: Listen & Learn.* London: Network Continuum

Sage, R. (2006) *The Communication Opportunity Group Scheme: Assessment & Teaching.* Leicester: University of Leicester

Sage, R. *et al.* (2010) *Meeting the Needs of Students from Diverse Backgrounds.* London: Continuum

Sage, R. (2017) (Ed.) *Paradoxes in Education.* Rotterdam: SENSE International

Sage, R. (2018) *One Funeral at a Time: Why We Hold on to Old Ideas. Journal of Linguistics. University of Buckingham Press.* Vol.10, pp. 43–53.

Sage, R., Rogers, J & Cwenar, S (2004). *An Evaluation of Education in England and Japan. The Dialogue, Innovation, Achievement and Learning Project.* Leicester: University of Leicester

Sage, R., Rogers, J. & Cwenar, S. (2006) *Why Do Japanese Children Outperform British Ones? The Dialogue, Innovation, Achievement and Learning Project.* Leicester: University of Leicester

San Diego City Schools (1982) *An Exemplary Approach to Bilingual*

Education. A Comprehensive Handbook for Implementing an Elementary-Level Spanish-English language Immersion Program, San Diego, CA: San Diego City Schools Publications, p. 183

Schmidt, R.W. & Frota, S.N. (1986) *Developing Basic Conversational Ability in a Second Language. A Case Study of an Adult Learner of Portuguese.* In R.R. Day (Ed.) *Talking to Learn: Conversation in Second Language Acquisition.* Rowley, MA: Newbury House

Seliger, H.W. (1983) *Learner Interaction in the Classroom and Its Effect on Language Acquisition.* In H.W. Seliger & M.H. Long (Eds.) *Classroom Oriented Research in Second Language Acquisition.* Rowley, MA: Newbury House

Schumann, J. (1975) *Affective Variables & the Problem Age in Second Language Acquisition. Language Learning*, 24, pp. 205–14

Sperry, R.W. (1968) *Hemispheric Disconnection and Unity in Conscious Awareness. American Psychologist*, 23, pp. 723–33

Vygotsky, L. (1962) *Thought and Language.* Cambridge, MA: MIT Press

Whitehead, M. (1997) *Language and Literacy.* 2nd Edition. London: Paul Chapman

Wilson-Draper, J. (2019). Details from the Director, Jeannie Wilson-Draper

5

EDUCATING HALF A BRAIN – UNRAVELLING CONCERNS ABOUT STANDARDS

'The brain is an enchanted loom where millions of flashing shuttles weave a dissolving pattern, always a meaningful pattern though never an abiding one.'

Sir Charles Sharrington

PREVIEW

The last chapter stressed the importance of using both our Left (L) and Right (R) brain sides to achieve effective performances. The media have joked about people with 'two brains', referring to their super-intelligence. The joke is also on us, as we all have two brains – a verbal and non-verbal one – with two ways of thinking. However, we exist on less than half our brain power and this issue is unpacked, unravelled and understood.

On the R side, we have one way of knowing. We 'see' imaginary things (mind's eye) or recall real ones. Imagine a favourite food – its colour, shape, taste and smell. We 'see' how things exist in space, understand metaphors, dream, fill in information/opinion gaps in talk/ text, combine ideas for new ones and assemble meaning (synthesise). If a thing is too complex to speak about, we gesture. Describe ringlets without hands! Images ('seeing' within) are idiosyncratic, non-verbal ways of thinking intuitively, holistically and metaphorically (Edwards, 1979). This is the 'seeing/feeling' brain, communicating with ourselves, understanding whole things/events.

The L side works oppositely. It analyses, abstracts, counts, marks time, plans in steps, making logical word statements and expressing thoughts. If apples are bigger than plums and plums bigger than cherries, we say that apples are bigger than cherries. This illustrates the L brain: analytic, sequential, symbolic, linear, objective and verbal. It is the 'saying/hearing' brain, communicating thoughts to others conventionally. Both brains complement one another but education favours L brain growth at the expense of the R. It leaves us full of facts but unable to apply them judiciously. Small changes in how we learn can produce big results, like having a 'tiger in our tank', as in an old petrol advert! Brain power is unleashed, so nothing seems impossible. We may even improve on our sixth place in the league table of history's most successful species! The lowly worm ranks first. Wriggling away, providing food for fish and birds, we underestimate the world's most adaptable creature. Charles Darwin, the evolutionist, argued that worms have had the greatest impact on civilisation because of an amazing ability to repair themselves, regenerate the earth and rescue us from environmental blunders (Lloyd, 2009). They are a feat of biological engineering. Learning to use both brains more effectively may shoot us up the evolutionary as well as the educational rankings for a double blessing!

INTRODUCTION: SETTING THE SCENE

Imagine a class of seven-year-olds. Heads are bent over hands busily colouring the Union Jack. Mrs Beaton (Mrs B), the teacher, spies a blonde staring into space:

'*Lissa, where's your crayons?*' She replies: '*Ain't got none, Miss.*' Mrs B sighs: '*I've told you not to say that. Listen: I do not have crayons; you do not have crayons; she does not have crayons; he does not have crayons.*' Lissa frowns: '*Miss... who's got all the crayons then?*' Here are two people with different thinking ways! Lissa (*using her R brain*) fixes on the whole context, whilst Mrs B (*using the L*) targets its components. Why is there no meeting of minds? At the age of seven, Lissa's R brain is gaining the gist of things, with her L one developing later to cope with details (Sage, 2000a).

This talk between teacher and pupil considers *what* they *say*, but is this similar to events we *see*? Look at Miller's (1986) puzzle below. What is it? A mess? Islands in the sea? Can you pick out a Mexican on a horse? Start with the middle top shape (*like a brimmed hat*) and work down. Unifying parts into a meaningful picture is the R brain role, whilst the L analyses components (*Mexican hat, male torso, horse's body, head, legs, tail, etc.*).

Miller's Puzzle Picture (1986)

We all have a double brain with different ways of processing and knowing – *seeing/feeling* and *saying/hearing*. The R considers the *whole* and the L *parts*. Examination shows how education concentrates on L brain tasks at the expense of the R, limiting thinking, understanding and judgement. Goldacre (2008) says the brain receives piecemeal information from many sources, but without a system to deal with it, the whole '*picture*' is not appreciated. He says constructing understanding is like looking at the vast ceiling of the Vatican's Sistine Chapel through a tube. With limited information and an inadequate R brain schema for making meaning, people are misled and unable to interpret evidence. This must be avoided to survive the hazards of a rapidly changing world. Pupils starting school today will enter jobs not yet existing and use technology not yet invented, so we need to carefully plan for their futures and develop all potential (Murphy, 2009). With 75,000 students playing truant daily, over half a million 11-year-olds not reaching a useful literacy level and 7 million working-age adults on benefits, we have challenges on our hands (Office for National Statistics, 2019).

GETTING TO KNOW BOTH SIDES OF YOUR BRAIN

The brain is a crinkled mass with unlimited power, resembling two walnut halves. It has an estimated 100 billion nerve cells and 500 trillion connections at the rate of 1 million per second. Knowing how power is harnessed to process and produce information is vital. Brains are functionally asymmetrical, with the L controlling the R side of the body and vice versa. However, language and its associated capacities are located in the L hemisphere for most of us. As speech and language express thinking and reasoning, scientists traditionally have regarded the *left* as the *major* hemisphere, with the right as the *minor* one. Roger Sperry's split-brain experiments changed this view. By cutting the *corpus callosum,* the nerve cable cross-connecting both hemispheres, Sperry isolated them to show differing functions.

Sperry's Split-brain Test (1968)

(Picture reproduced from Leicester course with permission for House of Commons lecture)

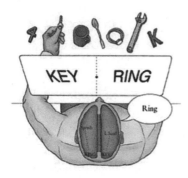

In a test, two different pictures flashed on a screen, with the split-brain man's eyes fixed on a midpoint to prevent scanning of both images. Each hemisphere received different pictures. A picture of a **key** on the L went to his R brain and a **ring** on the R to the L one, as above. When naming what flashed on the screen, he said **ring**. However, when reaching behind a screen with his L hand (*R hemisphere*) to select what was seen, he picked up the **key**. The L brain **says** and the R **sees**, but you cannot *say* what you *see* or vice versa if both are disconnected. For tasks like reading and writing you map what you *see* on to what you *say*, so both brains must cooperate. In activities like drawing, the hemispheres may work singly with the R brain *on* and the L more or less *off*.

Bias against the Right Brain in Culture and Tradition

Words expressing L/R concepts are frequent in language and thinking. The R hand (*L hemisphere*) is connected with *good, just* and *moral*. The L hand (*R hemisphere*) is linked to *bad, dangerous and immoral*. Bias against L hand/R hemisphere led parents and teachers of L-handed children to force them to use their R one for eating, writing and drawing, which sometimes led to learning problems. My L-handed mother had to change to her R hand at school. In my case, as a L-hander, I was allowed to develop naturally, but learnt to use my R one after a serious accident. In general, world languages show *good* connotations for the R hand/L hemisphere and *bad* ones for the L/R one. The Latin for L is *sinister*, also meaning 'bad', and for R: *dexter*, indicating 'skill' or 'adroitness'. In French, L is '*gauche*', or 'awkward', whereas R is '*droite*', 'just' or 'proper'. In English, L comes from the Anglo-Saxon '*lyft*', meaning '*weak*' or '*worthless*', whereas the R ('*riht*') is '*right*' or '*just*'. In politics, R-wingers endorse national power and resist change, whilst the L value individual autonomy and radicalism.

The R is *fascist* and the L *anarchist* at extremes. Regarding cultural customs, at a formal dinner the place of honour is on a host's R side. A bridegroom stands on the R at a wedding with the bride on the L – signalling relative status. Such references emerged when languages evolved in human L brains.

TWO WAYS OF KNOWING AND UNDERSTANDING

The duality of human nature and thought has long been debated by philosophers, scientists and teachers across cultures. Divisions have been made between *thinking* and *feeling, intellect* and *intuition, objective analysis* and *subjective insight* (Edwards, 1979, Sage, 2000a). Political writers say that people analyse good and bad points of an issue but vote with their gut showing that feelings override facts. Science abounds with anecdotes about solving a problem with an answer presented metaphorically in a dream. The 19th-century mathematician Henri Poincaré describes this: '*One evening I drank black coffee and could not sleep. Ideas rose in crowds. I felt them collide until pairs interlocked*' (Edwards, 1979, p. 35). We experience this when suggesting about someone: '*the words they say are fine, but something tells me not to trust them*'. Intuition shows that both brain-sides are at work, processing the same information differently. This implies that the R brain thinks in a free way, whereas the L has a fixed approach. The distinction becomes important when we consider how to solve a problem.

EUREKA: THE 'AH HA' OF UNDERSTANDING

In R brain information processing we use intuition and insight without figuring things out logically. The classic example is '*Eureka!*' (*I found it!*), attributed to Archimedes and his flash of insight while bathing, enabling him to form the principle of using the weight of displaced water to determine that of solid objects. Thus, the R brain mode is intuitive, subjective, relational, holistic and time-free. It is given short shrift in learning. Education cultivates the verbal, rational L hemisphere, leaving half the brain of each student virtually neglected (Sperry, 1982).

A WHOLE BRAIN IS BETTER THAN HALF OF ONE

With sequenced teaching based on words and numbers, schools are unequipped to develop the R brain role for creative thinking. Fixing only on event outlines and poor at categorising information, it focuses on the *'thing'* at present. We are uncomfortable with this free-thinking brain and more at ease with the L side's disciplined approach.

Educators acknowledge the importance of intuitive, creative thought, but schools continue to be structured in L brain mode. Teaching is sequenced; students progress through the years in a linear direction. The main subjects studied are verbal and numerical. Strict time schedules are followed. Seats are in rows. Learners produce answers – judged *right* or *wrong* – and are graded for standard government targets. The R brain (*artist, dreamer, free spirit*) is lost in our education system and mostly untaught. There are limited art, music and drama lessons, but we are unlikely to find courses in imagination, perception, intuition, inventiveness or communication. Educators and employers value such abilities, assuming students will spontaneously develop them.

Creative development occurs in spite of the school system, as survival depends on it. Does our culture so strongly reward L brain performance that we are losing potential of the other half? Levy (1970) said that our education system may eventually destroy the R brain with prescriptive methods. He was writing in an era with less emphasis on targets and perhaps more creative school experiences. We are aware of inadequate teaching of verbal abilities (*narrowly interpreted as vocabulary and grammar*) that handicap people for life. What happens to the non-verbal R brain (*assembling meaning*), which is barely trained? Neuroscientists provide evidence for R brain teaching, so we can develop education facilitating the *whole* brain, matched to natural development. R brain processing has its growth spurt between the ages of four and seven, with the L kicking in after this. Learning problems arise from a limited R brain strategy, as an early L analytic focus hampers growth. The result is a strong grasp of facts but a weak one for meaning (Sage, 2000a, 2003, 2010, 2017). It is significant that high-achieving countries start formal learning after the age of six, so working with the brain's natural development and allowing the necessary freedom for R brain growth and more real experience.

IMAGING IS VITAL FOR UNDERSTANDING

There is much evidence to suggest that those of us poor at imaging (*right brain activity*) struggle with understanding (Sage, 2003, 2010, 2017). All connected talk or text has information/opinion gaps which are dealt with imaginatively to achieve meaning. Seldom do we introduce *imaging* into class routines. How could it be done? In a lesson on Henry VIII, King of England from 1509, we might introduce him like this: '*Can you tell your neighbour what you think King Henry looked like? Sketch ideas and present to the class in five minutes.*'

This introduction helps students use their *imagination* so that when presented with facts about Henry VIII they will have been encouraged to use both brains. After *facts*, introduce a *review*. Ask student pairs to talk through what they gained from the lesson to then share with the class in a 30-second time slot each. Summarising experience develops *understanding* and a mental record. Without this opportunity only superficial learning occurs. We are driven to fill brains with *knowledge*, so marginalising approaches to help *comprehension*.

COUP D'ŒIL: ABILITY TO 'THIN-SLICE'

The phrase, *coup d'œil*, from the French, means '*power of the glance*'. Generals have this ability to immediately see and make sense of a battlefield. Each discipline has a word to describe the gift of reading deeply into narrow slices of experience. The tennis player who takes in and comprehends everything happening in a game has '*court sense*'. The ornithologist David Sibley says that in Cape May, New Jersey, he spotted a flying bird from over 70 metres away and knew it was a ruff. He had not seen this rare sandpiper before but captured what birdwatchers call the bird's *giss* – its essence (Gladwell, 2006). If not '*thin-slicing*' we would be unable to make sense of complex situations in a flash with everything seeming indecipherable. This R brain ability brings full understanding of events. This is facilitated by regular reviews and summaries of experiences. The R brain provides insight and event outline, leaving the L to analyse and arrange the parts. The Greeks grasped this in the *Places Trick*. Try it below to see how it works:

With eyes closed to block stimuli, imagine the front door of your house/flat. Open the door and enter the hall. Look around and see what is there. Now go into your kitchen and observe the space. Then, walk to the stairs/corridor for the bathroom. Step in and view. Finally move to your bedroom and enjoy gazing around this place of rest!

We have located six places: *front door, hallway, kitchen, stairs/corridor, bathroom* and *bedroom*. I will *tell* a story and ask you to recall the objects mentioned. Let us begin!

You have had a great night out with friends. As you roll home you find the front door ajar – propped open by a large red apple, the size of a football. This surprises you because the door was locked when leaving. You creep inside and on a hall table is a glass slipper, winking from the outside moonlight. It looks like the one Cinderella wore for the Prince's ball! Switching on the light, you troop to the kitchen, and on the workbench is a purple, sparkly pen, dancing an Irish jig! Life is getting curious! You rush to the stairs/corridor, finding a semi-precious stone, in your favourite colour, but the size of a large rock. You step over this carefully and rush to the bathroom. As you enter, the smell of your favourite oil greets you, and luxuriously bathing is a large wooden spoon, which turns and grins! Backing out, you find the bedroom, and on your bed a velvet cushion, with a silver necklace curled like a snake!

Without looking back at the text, can you recall the items? How was this achieved? Did you remember the six familiar places and link these to the new information? It is the R brain that creates these associations. Descriptive words help to imagine the scene. Knowledge of language grammar, with its syntactic arrangements, assists the organisation of detail. It shows how both brains work to carry out tasks effectively, thinking both inductively and deductively. Putting together what you *see, feel, say and hear* is the essence of the process. Both brains complement rather than contradict, in order to give us breadth and depth of understanding. This brings us to consider how we shift brain modes for learning using the *communication* system as the essential tool.

THE SHIFT FROM INFORMAL TO FORMAL TALK OR WRITING

Development depends on opportunities to structure thinking from formal talk

Informal talk is unplanned and unstructured **Meaning** comes from context	*Formal talk or writing* is planned with a beginning, middle and end **Meaning** comes from words and filling in information/opinion gaps
Rosie: '*Jack's greedier than Russell.*' **Helen:** '*Um, cos he's male.*'	'*Rosie and her daughter Helen are at the back door feeding their Indian Runner ducks, Jack and Russell. Jack dives in aggressively, devouring the bread before Russell, the female, gets a look in.*'

In *informal talk,* meaning depends on *seeing, observing and feeling* what is happening in the *context*. In *formal talk/text*, descriptive words represent reality, but gaps in information use imagination for understanding. We do not know the ages of Rosie and Helen, so must guess. Mention of Helen, as the daughter, gives us a clue, but we have no means of knowing whether she is a child or an adult. The fact that Russell (*male name*) is female needs resolving, as it depends on a story of which we are unaware. Rosie's husband wanted a *Jack Russell* dog, but travelled the world so was not around to be responsible. A compromise was made with two ducks, *Jack* and *Russell*. Most narratives have hidden information which the R brain resolves, using lateral thinking and *inference, reference* and *coherence*, to produce a meaning likely to be different for each of us.

TEACHING THAT IS WHOLE-BRAINED

Thinking starts with the R brain creating a range of possibilities and selecting the best fit. The L brain then sequences steps to the goal as in the picture below. Bruner (1966) and Beilin (1975) emphasised that R brain thinking is the primary mode for problem-solving, but preoccupation with the fixed framework of logical development (L brain activity) means it receives less attention. Logic is important in reaching conclusions but not infallible. Take the following statements:

Birds fly. Penguins are birds so penguins fly.

Penguins are exceptions to the rule and do not fly. Even flying birds cannot do so if legs are stuck in mud or wings are damaged. It is the R brain that thinks *laterally* and *creatively* to connect up a range of information from many sources for considering the whole event. It operates frecly, following any direction and begins problem-solving.

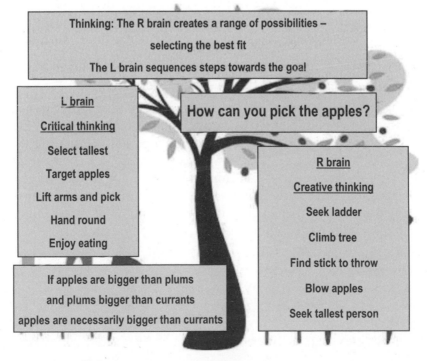

Thinking: The R brain creates a range of possibilities – selecting the best fit
The L brain sequences steps towards the goal!

L brain

Critical thinking

Select tallest

Target apples

Lift arms and pick

Hand round

Enjoy eating

If apples are bigger than plums and plums bigger than currants apples are necessarily bigger than currants

How can you pick the apples?

R brain

Creative thinking

Seek ladder

Climb tree

Find stick to throw

Blow apples

Seek tallest person

Take: *How to pick apples from a tree*. You start by generating a range of ideas (*listed right*) and after selecting the '*best fit*', move to the L to work out logical steps to the goal. Thinking becomes narrow, with less possibilities, if the R brain is inactive. R brain thinking begins the process and brings together the whole experience. It is known as *narrative thinking,* seen in children when they '*story*' real experiences in play, like a visit to the shops, etc. This thinking is often ignored and not clinically tested.

NARRATIVE THINKING AND STRUCTURE

I was asked by the Medical Research Council to look at children thought to be intelligent but underachieving in school. Tests of *thinking, understanding and expression* rarely require subjects to assemble quantities of information so that *narrative thinking and structure* are not monitored, but presented as major learning problems (Sage, 2000a, 2017). Such children may function well in dialogue where there is equal participation, but flounder when faced with monologue situations, where one controls actions, because of their problems with bringing together information for overall meaning (see Appendix).

Narratives develop when children recall and retell events, give instructions, explanations and reports. Today, *television* replaces *formal talk* in life. When you watch a programme, meaning is provided in ready-made images, with talk only a secondary back-up. Viewers are passive and both brains do not have to actively think and visualise to understand. Modern life gives no time for talk – key to imaginative development of ideas.

Children, starting school, are shocked because suddenly they are confronted with vast amounts of talk with no mental schema for dealing with this. Studies indicate that many begin education with thinking and language well below that required to cope with formal class talk (Sage, 2000a, 2017). Cwenar (2005) showed that 80 per cent of students entering her senior school had thinking and language ages of around five years at age 11. Early deficits had not been made good because no sustained attention was paid them in primary schools. They are remedied by attention to formal communication, focusing on how ideas develop into narratives in the *Communication Opportunity Group Strategy* (Sage, 2000a) below. Researchers attest to its success (Cooper, 2004).

The approach shows how ideas are generated and organised progressively. From producing a range of ideas at the first level, you can eventually locate context, characters, events, results and reactions. Evidence suggests that many students leave school without achieving

A MODEL TO ASSEMBLE INFORMATION AND OPINION (Sage)
7 stages to information development

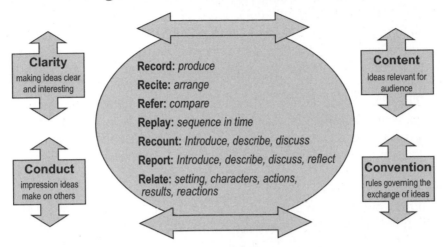

Clarity
making ideas clear and interesting

Content
ideas relevant for audience

Conduct
impression ideas make on others

Convention
rules governing the exchange of ideas

Record: *produce*

Recite: *arrange*

Refer: *compare*

Replay: *sequence in time*

Recount: *Introduce, describe, discuss*

Report: *Introduce, describe, discuss, reflect*

Relate: *setting, characters, actions, results, reactions*

Supports personal and academic development

the last three levels, so limiting thinking, reflection and judgment (Sage, 2000b, 2004, 2017). Development cannot be left to chance, with *formal talk* enabling both brains to function equally with the chance to *see, feel, say and hear. The Communication Opportunity Group Strategy* works for groups of all sizes, and focuses on narrative thinking and structure, along with making ideas clear **(clarity)**, ensuring they are audience-appropriate **(content)**, whilst obeying social and linguistic rules **(convention)** and behaving in a proper manner **(conduct)**. Research supports this, bringing both academic and social success (Sage, 2000a, 2000b, 2003, 2004, 2006, 2017). However, 200 years ago, thinking was different, with Lancaster (1806) writing: '*Talk in school is very improper – children cannot talk and learn at the same time. In my school talking is considered an offence and an appropriate punishment succeeds.*' Does the view still linger?

THE FUTURE

Institutions and individuals find it difficult to change and favour accepted doctrine and routines because of tradition. There is public clamour for better academic standards, against government claims that they are rising. It is confusing and overwhelming. Dispute rages as to whether education matches present needs, with the Confederation of British Industry reporting employee problems with spoken and written communication that seriously affects performance. Tesco's Chief Executive has bemoaned the '*woefully low*' education standards, saying in the media: '*Employers like us are often left to pick up the pieces.*' Others dispute this view, suggesting the population has never been better educated. This might be so for *knowledge*, but workplaces favour *personal competencies,* dependent on R brain development, with the ability to apply what is learnt, size up events and respond appropriately, accounting for other feelings and views.

The '*standards*' debate is illuminated by other nations' experiences. Japan is an example. Years ago their economic bubble burst and people could not cope. Experts agreed that a strong, traditional academic education focus had been at the expense of *transferable abilities* that enable survival. Now, *communication and relationships* take priority in schools, on the basis that *talk* is the technology for understanding. Talking together achieves clarification and comprehension of ideas which Socrates (*Greek philosopher*) called *maieutics*, or *intellectual midwifery*. Effective communication not only brings about better understanding but forms social contact and develops cooperation and collaboration. Japan has seen benefits in handling globalisation more effectively than other nations when measured by international indicators. The Japanese concentrate on helping students to understand through talk, making them aware of the role of non-verbal aspects, like voice, gesture and facial expression, as markers for meaning. Education for humanity and understanding is at the core of their philosophy.

The Dialogue, Innovation, Achievement and Learning (DIAL) projects enabled English and Japanese teams to share expertise in pursuit of the ideal global citizen (Sage, Rogers & Cwenar, 2004, 2006, 2010). Although the Japanese feel they lack creativity, I am not so sure! Professor Tsuji of Tokyo University has grown replacement teeth for mice, so dentures should be a thing of the past. He thinks they will develop fully functioning bio-engineered organs to replace damaged ones, just like the earthworm achieves. You cannot get more creative than that! Cuba is another nation at the top of UNESCO tables of achievement but cannot afford books. To Cubans, *talk* is the way to share, mediate and extend learning, and abandoning students to book study prevents this, restricting thinking and understanding. Therefore, it can truly be said to know one's own country one must know another one for comparing performance and considering improvements.

NO CONSENSUS ON EDUCATIONAL GOALS

As a nation we have no clear consensus on what education should give us. What kind of human being do we want to be? A wish should be for us all to become educated in things perfecting our nature – *humanity, knowledge and understanding*. Most agree that our present system targets the middle attribute with *humanity* and *understanding* more debatable achievements. Education must be concerned with the real values of life, helping us to live wisely, agreeably and well. In an illuminating article, 'A Talent for Dishonesty' (2009), Farndale refers to a student poll from a top university where 50% admitted to cheating: '*I Google the essay title, copy and throw everything on to a blank word document,*' said one. Supporting this idea, the following appeared in my email inbox:

'*To buy a degree is easy these days. Nevertheless, most students just sit around in their boring University classes, wasting money. Why would you do that? Buying a degree is a matter of personal motivation, but why should you buy a degree? The main reason is that buying an online degree is going to save you a lot of time, a lot of work and will make you a lot of money. It is the intelligent solution.*'

How does one respond to that! A popular book for young people is *The 48 Laws of Power*, which advises '*Pose as a friend, work as a spy and crush your enemies totally*' (Green & Elffes, 2000). The message is simple. It is fine to cheat because it is now OK to lie. '*Spinning*' is the new word for '*lying*', and common practice, but a road to disaster. Audur Capital was the only institution to survive the Icelandic bank catastrophe because of high moral principles: *risk-awareness; profit with principles; emotional capital; straight talking and independence*. It is a potent formula, reflecting L and R thinking and a broad, deep understanding of humanity and respect for others. The Glendinning Report (2019) confirms how cheating by students and staff is rife across the world.

THE EDUCATIONAL SYSTEM

The phrase '*educational system*' is self-contradictory, as the learning aim is to develop human potential and not fit them to a standard framework. Education reflects what politicians want it to do, and change only happens if the community supports this. Some believe that attempts to educate everyone must end in the education of nobody. They assert that differences between us are so great and confusing that dilution of a common study programme is inevitable. There is truth in this, and Cuba's Universal Policy has much to commend it. This respects that everyone should be educated, but in line with interests and ability and in accordance with society needs for a skill range. Cubans observe that in Britain we have made an *elite* education, developed originally for those entering traditional professions, the destiny of all, when the bulk of us will have practical careers, dependent on a broader knowledge and skill base (Sage, 2009). Goldacre (2008) supports this, saying that obsession '*with taking the message to everyone rarely offers stimulating content to those who are interested*' (p. 321).

Paradoxes in Education (Sage *et al.*, 2017) presents views of different stakeholders in education to reflect on these issues. The unanimous conclusion of contributors is that effective education depends on us understanding and respecting one another, with communication as key to this. If it aims at *understanding* it will make the most impressive contribution to power and prosperity, but if targeting just the latter, this ambition will fail. A brain is not a receptacle, and education is what remains after facts are forgotten – the ideas, strategies and habits that enable us to survive and cooperate with others. Learning does not happen unless teachers and students achieve a *meeting of minds* and can communicate together. As a learning society largely of our making and in turn making us, institutions have been designed to perpetuate existing principles and practices. With the speed of change ever accelerating, we are forced to reconsider present values and redirection of education towards new ones. First steps are a better understanding of the real facts of life, the potential of both brains, the new values to be attained, the communicative competencies required amongst diverse cultures and the possibilities and limitations of education in achieving them.

Rene Dubos, the geneticist, in *Man, Medicine and Environment* (1968) demonstrated that we use less than 20% of our brain potential. Decades later, evidence suggests we may not have improved on this. There are people whom and places where we can look to for help. Inspiring schools put communication and relationships at the core of their enquiring minds philosophy. By co-ordinating R and L brain activity through formal talk experiences we have the key to superior performances and the best chance of developing students' humanity and understanding as well as knowledge. Those reading this will know of examples where power, passion and possibilities have been released. We must spread the word, helping everybody to have a better understanding to assist students to be *whole-* rather than *half-brained*. Next time you spot a wriggly worm in a garden, reflect on these remarkable creatures and their extraordinary ability to regenerate, adapt and survive successfully. They have a message we cannot ignore.

Review

Karl Friston's (2009) work on a mathematical law of the brain attracted interest and may mean we will eventually have a brief, iconic equation like Einstein's $E = mc^2$ to flash on T-shirts. It suggests a new metaphor of self-identity focusing down on the *'free energy'* in the brain system and the meanings relevant to a particular situation. When walking along, our senses inform us about surroundings to avoid falling over or bumping into things, with our brains working to help us avoid surprises like holes in the ground. New theories will change our concepts of how brains work and help us understand even more the deeper mechanisms of mental performance. Recently, the work of Barbara Demeneix, a biologist from the Sorbonne, Paris, has alerted us to the dangers of chemical pollution on brain development. She discusses the effects of chemicals that disrupt the thyroid hormone. This is a crucial hormone needed for optimal brain development both before and after birth. She synthesised the results from epidemiological and experimental studies linking the increasing numbers of chemicals affecting thyroid signalling with IQ loss and/or neurodevelopmental disease. This work is largely based on recent research that forms the foundation for her latest book – *Toxic Cocktail* (2017). Meanwhile, children need *'guides by sides'* rather than *'sages on stages'*, with the freedom to gather a pocket full of dreams and a focus to make them a reality. Greater development of the R brain is essential now robots are able to carry out much L brain routine analytic activity. The freer, imaginative R brain is the key to our future survival.

MAIN POINTS

1. We all have a seeing R brain and a saying L brain, which must work together for effective learning, and is coordinated through talk activities.
2. The L brain processes analytically and serially and the R one holistically and spatially.
3. Experts say that education neglects the R brain and favours the L, to restrict free-thinking, creativity and the imagination required for understanding.
4. There is no consensus on education goals with more attention to academic rather than moral standards.
5. Teaching to standards means we often neglect potential.

REFERENCES

Beilin, H. (1975) *Studies in the Cognitive Basis of Language Development*. New York: Academic Press

Bruner, J.S. (1966) *Toward a Theory of Instruction*. New York: Newton

Cooper, P. (2004) *Successful Approaches with SEBD Pupils*. London: Dr Barnardos

Cwenar, S. (2005) Unpub. PhD from an Educational Action Zone project to audit student abilities. Leicester: University of Leicester

Demeneix, B. (2017) *Toxic Cocktail*. New York: Oxford University Press USA

Dubos, R. (1968) *Man, Medicine and Environment*. New York: Frederick A. Praeger

Edwards, B. (1979) *Drawing on the Right Side of the Brain*. London: Fontana

Friston, K. (2009) *The Free Energy Principle: A Rough Guide to the Brain. Trends in Cognitive Sciences*, 4 (11)

Gladwell, M. (2006) *Blink: The Power of Thinking without Thinking*. London: Penguin

Glendinning, I., Maris-Orim, S. & King, A. (2019) *Policies and Action of Accreditation and Quality Assurance Bodies to Counter Corruption in Higher Education*. Pub. by the Council for Higher Education Accreditation

Goldacre, B. (2008) *Bad Science*. London: Fourth Estate, HarperCollins

The Graduate Recruiters Association (2009) *Employers Survey*. London: GRA

Green, R. & Elffes, J. (2000) *The 48 Laws of Power*. London: Penguin

Lancaster, J. (1806) *Improvements in Education*. In Adrian Jackson (Ed.) (1993) *Desperate Deeds: Tales of Triumph and Despair*. London: Collins Classic Collection

Levy, J. (1970) Information processing and higher psychological functions in the disconnected hemispheres of commissurotomy patients. Unpublished doctoral dissertation, California Institute of Technology (Ann Arbor, MI: University Microfilms No. 70-14, 844).

Lloyd, C (2009) *What On Earth Evolved? 100 Species That Changed the World*. London: Bloomsbury

Miller, L (1986) *Language Disabilities, Organizational Strategies and Classroom Learning*. Research paper: The Language Learning Disabilities Institute. Emerson College, San Diego, CA

Murphy, G (2009) *A Vision of Learning for the Future:* Presentation to the Inter-competency & Dialogue through Literature (IDIAL) project with Bulgaria, Finland, Latvia, Slovenia, Spain & the United Kingdom: Liverpool Hope University

National Statistics (2009) Data published by the Department of Children, Schools and Families Sept. 2008–April 2009

Sage, R. (2000a) *The Communication Opportunity Group Scheme*. Leicester: University of Leicester

Sage, R (2000b) *Class Talk: Successful Learning through Effective Communication*. London: Network Continuum

Sage, R. (2003) *Lend Us Your Ears: Listen and Learn*. London: Network Continuum

Sage, R. (2009) *What Cuba Has to Teach Us about Education*. *Education Today*, 59 (3)

Sage, R., Rogers, J. and Cwenar, S. (2004) *An Evaluation of Education in England and Japan*. The Dialogue, Innovation, Achievement and Learning Project. Leicester: University of Leicester

Sage, R. (2017) (Ed.) *Paradoxes in Education*. Rotterdam: SENSE International

Sperry, R.W. (1968) *Hemispheric Disconnection and Unity in Conscious Awareness*. *American Journal of Psychology*, 23: pp. 723–33

APPENDIX: DIALOGUE AND MONOLOGUE

The word *dialogic* has two related root words: *dialogue* and *logic*. Many experts have written on dialogic theory, but Martin Buber's ideas have become so important that many use his terminology. Buber's book, *I and Thou*, suggests that a truly humanistic relationship requires both parties to come to the relationship without preconditions, fully accepting the other. This definition emphasises acceptance in the relationship.

Monologic communication
In *I and Thou*, Buber notes that much of what we consider communication is not about people *'an und für sich'* (*in and of itself*), but about functions. These relationships are between *I* and *it* (*I-it* relationships), working when functional – a doctor diagnoses a patient as an *it*, because objectivity is desirable in that context. However, Buber says this view can become habitual, and the *I-it* relationship structure suffuses all relationships.

The I-Thou relationship
In the *'I-thou'* relationship, Buber says we make ourselves fully available to the other person. We are open, willing to share ourselves completely and want to understand another world-view. When monologic communication takes the place of dialogic, we become incapable of experiencing others and ultimately alienated from our own experience. With dialogue, we achieve real communion with others and experience them in what Buber's followers have called *'the now'* or *'the here and now'*. This is, to use Jean-Paul Sartre's

terminology, an *existential relationship*. It is not predefined, nor can it be constructed. It occurs when open to experience.

THE EXISTENTIAL I-THOU RELATIONSHIP

Buber states the impossibility of achieving an existential *I-thou* relationship through communication strategies or other rational/rationalist attempts. He was deeply religious, believing that experience is there for us when we accept it, often in the form of a religious one. For Buber, God is the ultimate *'thou'*. When we give up trying to prove God's existence – a rationalist attempt, reducing God to an object – we can then accept the reality of God. Likewise, when we cease trying to construct relationships with others and simply accept the other person in the *here and now*, without definition or limitation, we have a truly dialogic experience, a shared one between *I* and *thou*. This, says a Buber follower, has five characteristics: it is mutually open, non-manipulative, confirming, non-evaluative and recognises the uniqueness of the other. The result and the sixth characteristic is that the two parties reach *a common fruitfulness* unique to them.

LIMITATIONS OF DIALOGIC THEORY

Academics have criticised Buber's theories for their vagueness. If you cannot work to achieve dialogue, how useful can the theory be? Public relations firms embrace the theory and use it to construct a synthetic dialogue between the corporation and its customers. The University of Memphis course notes on the theory cite the famous ad tag *'Reach out and Touch Someone'* as an example of this – a synthetic approximation of real experience. Twitter, with its emphasis on 280 characters or less, is another synthetic approximation of authentic communication.

6

BRIDGING EDUCATION AND WORK – ISSUES FOR TEACHER TRAINING AND PRACTICE

PREVIEW

There is a mismatch between school and work requirements and a sense that what we have been led to believe about education and qualifications is somehow false. Research has found an inverse correlation between qualifications and job performance. Education has been viewed as good for society, but tends to have a corrupting effect on genuine learning – concentrating on its products (examination results) at the expense of processes (application of understanding). Job recruiters look for psychological and social aptitudes that are not easily defined or measured. These develop from attention to individual needs and key competencies which have been sidelined in pursuit of test results and school rankings. Many teachers, however, view their job as not just pointing students to qualifications as a means of livelihood but to a broader view of what a fulfilling life looks like. This requires rejecting a narrow academic life course that is regarded as obligatory and inevitable in British and many other societies! A focus on developing attributes and virtues required for successful living and working is advocated. Teachers play a vital role in nurturing these and must give them priority in learning.

THE CONTEXT

When visiting Cuba, I arrived at a school in Havana to find the playground full of gorgeously dressed boys and girls practising a dance for a carnival. The dance master was a student from the group who led them through all the movement sequences. Afterwards they went into a mathematics lesson and plotted the dance moves to examine their patterns. Later, in a language and communication class, students enacted a fashion parade in their costumes, with a Master of Ceremonies (*a pupil*) who introduced each one, posing questions and asking them to explain how their outfit was made. Learning was very '*hands-on*' and rooted in real experience, with students constantly encouraged in the practical arts and demonstrating their self-reliance, thinking and communication, as well as control over the process of learning in participatory ways.

In contrast, on visiting an English school, I found similar-aged children of 8–9 years confined mainly to classrooms and absorbed in worksheets, books and exercises for most of the day. Real experience was abandoned in pursuit of the acquisition of abstract knowledge, resulting in a dependence on adults who delivered this in prescribed ways. Students were not regularly called upon to exercise practical competencies or self-reliance in a participatory approach. Some people would even view it as irresponsible to educate the young in an active way for practical futures, which are regarded as relics of past history in a world where we press buttons to get things done.

Teachers often feel impelled by impersonal forces from outside the classroom. However, there is rising concern that we are becoming less able to gain an adequate grasp of an increasingly complex world. This can only be achieved by getting hands-on, real-life experiences to develop full understanding. In tough economic times we must learn to be frugal, which requires self-reliance, self-control and the ability to make do and survive. There is need for a vision that is human in scale, with less dependence on forces of the global economy over which we seem to have little control. Expectations of the 'perfect life', with pots of money to make this happen, need revising!

Schools have to prepare students for a world of work where individual responsibility is compromised by a bureaucracy which arbitrarily dispenses credit and blame. The rise of teamwork makes it impossible to trace *who* is responsible for *what* and allows managers to manipulate workers. Managers, in turn, are stressed by imperatives to meet impossible targets. Those leaving school and seeking jobs find recruiters are less interested in academic grades and more in their personality and ability to communicate and cooperate. Is educational achievement, therefore, a sham? There evidently is a mismatch between school and work requirements and a sense that what we have been led to believe about education and qualifications is somehow false. Many teachers, however, view their job as not just pointing students to a means of livelihood but to a broader view of a fulfilling life. Turner (2013) suggests that schools providing *'feasts'* (*a personal as well as academic diet*) rather than *'basic rations'* (*academic diet only*) produce students capable of living more satisfying lives.

From confusion of ideals it may dawn that *productive* work is the foundation of prosperity and success in any society. Abrahams (2013) suggests that, in Britain, 60% of university graduates are in jobs for which a degree is irrelevant, such as serving drinks in a bar or stacking shelves in a supermarket. There is a chronic shortage in the traditional trades (*e.g. plumber, electrician or carpenter*) with a recent *Wall Street Journal* article saying they are becoming one of the few sure paths to a good living, because they are meaningful, useful and can be well paid hour-for-hour. The experience of making and fixing things has been disappearing from education, as we concentrate on acquiring facts for standard testing and qualifications. It is time to remedy the problem, and this begins in schools, which exist to assist personal growth and ensure economic independence. We have a generation of students who can answer questions on standard assessments but are less good at doing what workplaces require.

WHAT ARE THE BENEFITS OF HANDS-ON LEARNING?

Alexandre Kojève (1989), the philosopher, suggests that by producing something we recognise and discover ourselves in the result. Through hands-on experience and the competencies gained we become *'quiet and easy'*. Pride in achievement of something tangible is far from the *'self-esteem'* that educators are keen to cultivate in students. Such ideas are extended by Arendt (1958, p. 23), who says that what we produce gives *'rise to the familiarity of the world, its customs and habits of intercourse between men and things as well as between men and men'*. With regard to material things, like machines, those who are able to make and repair them when broken put themselves in the service of others, observing, listening and noticing, in order to solve problems and diagnose faults. This challenges the understanding of those unable to cope when normal things are disrupted in their lives, because they lack the ability to design, make and repair objects. I constantly come across young people who cannot sew a button on to clothes – which amounted to a *'sin'* when I was younger!

Political theorists, however, like Aristotle, questioned the virtues of artisans, finding them too narrow in their concerns for public good. This was before mass communication and conformity eroded personal independence, judgement and initiative. Plato, moreover, differentiates *technical competence* and *rhetoric* on the grounds that the latter *'has no account to give of the real nature of things, and so cannot tell the cause of any of them'*.

In British schools, *hands-on* activities are given limited attention and value because by acquiring a specific skillset one's life will be determined and constrained. In contrast, a university education is regarded as a means to an open future, although the move to functional degrees, like *beauty therapy* and *computer gaming*, could be considered as equally restrictive. Craftsmanship requires learning to do one thing well, to a recognised standard, whereas today we value learning arbitrary facts, embracing potential rather than achievement. Work patterns often require employees to be responsible for only small parts of a whole process, with less understanding of what is totally involved. This can weaken personal accountability, motivation and work commitment.

The pride in being good at something and seeing things through from start to finish results from accumulated experience and time spent in achieving excellence. Today, quick results are required and the ethos and ethics of craftsmanship are not respected. Recently, I met someone at a social event who had given up work as a carpenter and joiner because it had no status, becoming instead a manager in a care home for improved social standing. Managers often lack any real expertise, presenting a commanding status and attractive lifestyle, which subordinate workers cannot match. Everyone is dazzled by the prospect of being a *knowledge worker* with a fat salary for *knowing* things, but not necessarily understanding or able to implement these practically. However, the dichotomy of *knowledge* versus *manual work* rests on misconceptions. Useful, hands-on work can be intellectually challenging and economically rewarding. I know someone with a PhD in genetics who now is working as a plumber, earning four times the salary he did as a researcher. Also, the man who sweeps the platform at our local railway station is earning £5,000 more than my son of the same age, who is a senior lecturer, with a PhD and MPhil in psychology, plus a first degree in medical sciences. Knowing a great deal does not always pay!

THE MENTAL CHALLENGES OF HANDS-ON EXPERIENCES

Rose (2005) has written about the cognitive demands of hands-on work. Skilled manual labour interfaces with the material world and is rooted in natural sciences. Knowledge of materials and tools, and how they behave in particular circumstances, is acquired through close perception and reflection in real situations. For Homer, this *skill* meant *wisdom*. Through repeated experience, a carpenter learns about the many types of wood, their fitness and strength for load-bearing, stability in different weathers, resistance to rot, porous features and susceptibility to insects, amongst other things. Practical work gives knowledge of universals like weights, measures, volumes, angles and levels vital for effective constructions. These underpin skills in design, cutting, assembly and finishing of parts into a final product. In fact, craft practices have given rise to scientific understanding and not vice versa. Take the steam engine, developed by mechanics Thomas Savery, Thomas Newcomen and James Watt, who observed the relations between volume, pressure and temperature. At this time, theoretical scientists were promoting a caloric theory of heat which proved to be a fruitless pursuit. This example illustrates Aristotle's point:

'Lack of experience diminishes our power of taking a comprehensive view of the admitted facts. Hence those who dwell in intimate association with nature and its phenomena are more able to lay down principles such as to admit of a wide and coherent development; whilst those whom devotion to abstract discussions has rendered unobservant of facts are too ready to dogmatise on the basis of a few observations.' (*On Generation and Corruption*, 316a 5–9)

Certain aspects are often better represented by hand than in language. I remember being able to make paper-serviette water lilies by watching my mother do this for parties, but years later I came across written instructions and found them too complicated to follow! Such examples help us to understand why certain aspects of production cannot be reduced to rule-following alone. Given the intrinsic satisfaction resulting from hands-on activities – mentally, socially and in a broader psychological appeal – one wonders why it is devalued as a leading component in our educational system. The view that manual work is disappearing with the rise of technology is questionable and unlikely. It is to history and culture that we must look for understanding how work with our hands has become debased in British society.

WHAT WE LEARN FROM HISTORY

In medieval times, the trades and their exquisite guild houses (*headquarters*) were the elite of the land. The emergence of industrial societies, however, led to more differentiated, routine work, and crafts faded into obscurity as factories dominated and mechanised assembly lines speeded up production and quickly built wealth for *managerial* classes. Many manual jobs were made redundant and trades lost their important place in society. The 20th century saw jobs radically simplified with the growing separation of planning from execution, due to automation and rational administration. Braverman (1974) talked about '*scientific management*' as '*management masquerading in the trappings of science*'. Managers have taken the traditional knowledge of workers and reduced it to rules, laws and formulae. Craft knowledge, learnt from doing the job, is given out to workers in limited instructions needed to perform a *part* of a whole *process*. This removes the brainwork from the job and places it in the strategy and planning sector. Nowadays, the thinking aspects of jobs are with administrators, with skilled workers easily replaced with unskilled ones at lower pay rates. This has contributed to a greater separation of the rich and the poor. Administrators trump professionals, which has led to ineffective performance and eventually company crashes, as in the recent Carillion example. Some years ago, they replaced engineering managers with accountants, who did not understand the business, so eventually it failed.

The introduction of *time and motion analysis* describes the physiological capacities of the human body in machine terms and defines *abstract labour* as in an assembly line. Workers reluctantly and eventually came to accept the assembly line because they were compensated by higher wages. With greater production, *consumption* also needed to be brought under scientific management – the *management of desire* – which establishes expanding needs in people to justify more output. This has led to consumers buying on credit and required to keep jobs at all costs to pay off debts, so making them unlikely to challenge the status quo.

THE DEGRADING OF WORK

Today's post-industrial economy has everyone dealing with abstractions. The academic professions (*medicine, education and law*) are now subject to routines and degradation, following the logic that hit the manual trades a century ago. The thinking elements of jobs are removed from professionals in a system devised by administrators. Teachers have a prescribed curriculum, judges receive guidelines for sentencing and doctors abide by regulations for drugs and operations. This means that knowledge work is not growing but actually shrinking and concentrated in a diminishing elite. This has huge implications for the vocational advice that students should be receiving from schools, as this situation conflicts with educational goals to fit greater numbers of students for the knowledge economy.

Garson (1989, p. 46), in *The Electronic Sweatshop*, describes '*how extraordinary human ingenuity has been used to eliminate the need for human ingenuity'*, with expert systems transferring *'knowledge, skill and decision making from employee to employer'*. Counteracting this, however, the *'creative economy'* has been promoted to encourage worker participation. Creativity is produced from long, dedicated, constant practice, and the new requirement for employee flexibility, to cope with rapid workplace changes, prevents concentration on a task long enough to develop real mastery and competence. We are expected to be masters of everything, but end up being masters of nothing. *The World's Broken Workplace* (Clifton, 2017) analyses a Gallup poll of 160 countries, finding that 85% of workers hate their jobs and bosses. Grafstein & Nelson (2019) in response say managers need three rules to change worker dissatisfaction:

1. **Culture.** Only 27% of employees have the same values as their employer, so emphasis must be on producing a culture acceptable to all.
2. **Engagement**. Only 15% of workers say they are engaged at work, so managers must focus on their development to bring more personal fulfilment.
3. **Problems**. 85% of managers cannot analyse data, but are obsessed with numbers, targets and money-making rather than solving worker and client problems.

Where does this leave the teacher expected to give guidance to students regarding their futures? Although manual and non-manual professions have been subject to routine systems over the past century, there are jobs resistant to such constraints. Take the work of those involved with building construction. Although prefabricated doors, window frames, staircases and roof trusses have eliminated some difficult job aspects, there are physical and individual circumstances of work performed by builders, plumbers, carpenters, electricians and mechanics, etc., that require skilled responses, allowing the feeling of being a person rather than a machine. Trades and crafts are increasingly the route for those who want to live and work by their own steam and through their own powers. The example of the PhD researcher who is now a plumber suggests this, as he wanted more autonomy, satisfaction and security from employment, which led him to abandon the lottery of bidding for project money to undertake research studies.

The advice to students should be that only if they have a bent for books and study should they decide to go to university. Pells (2017) reports that 65% of students are dissatisfied with their studies. University of Buckingham is top of the list for student satisfaction with their education (97%), because lecturers are passionate about their subjects and small classes mean they have individual attention, which larger institutions cannot provide. Again, small is beautiful! Academics must be encouraged to approach their work with the ethos and ethics of a craftsman, delving deeply into the sciences and arts and developing real subject expertise and competence. Society needs these people, as they are able to bring together bodies of knowledge and from this synthesis advise others on how to improve world circumstances. If not inclined to academic study, students should be encouraged to pursue a trade or craft, as it will be likely to provide greater personal satisfaction, security and possibly better pay as an independent worker rather than a corporate, computer-hugging operative. Such practical experts are hard to find in Britain, but are essential to the smooth running of society. This requires rejecting an academic life course that is regarded as obligatory and inevitable in British society! The practical intelligences must be better valued and given the important status they deserve. It is a positive step forward that the present UK government is supporting the move to introduce more apprenticeship schemes across professions.

A QUESTION OF AUTONOMY

People are brought up to think that the world has no limits and they can master anything if having the will to do so. However, self-realisation and freedom are about buying new things rather than producing our own or conserving those we possess. When thinking about hands-on engagement, we must consider human nature, in order to illuminate the practice of constructing, fixing and routinely maintaining things. This requires discipline and obedience to the reality of a task. Take the experience of learning a foreign language. Iris Murdoch (2001) describes this in *The Sovereignty of Good*, p. 21:

'*If I am learning, for instance, Russian, I am confronted by an authoritative structure which commands my respect. The task is difficult and the goal distant and perhaps never entirely attainable. My work is a progressive revelation of something which exists independently of me...*'

In any difficult discipline one submits to their ways, but consumerism has a different conception of reality. Crawford (2009) distinguishes between *commanding* and *disposable* reality, corresponding to *things* versus *devices*. The former brings meaning through inherent qualities, whilst the latter answers to our shifting psychological needs. People play music less than they did and listen to music on their devices instead. An instrument (thing) is hard to master and limited in range, whereas a device makes all music instantly available, bringing personal autonomy. A *thing* requires active, skilled, disciplined engagement, while a *device* invites disposable consumption in a throw-away society that now builds things with short lives to promote sales. The modern person is being reprogrammed to become a passive consumer, starting with computers and games consoles, monopolising time. Children easily become removed from the activity of construction, evaluation, reflection and review, as well as the communication and cooperation that result from this engagement.

EDUCATION FOR DIVERSITY

Consider a dressmaker and a dog-trainer. One has to be careful in cutting, sewing and fitting a garment, and the other constantly communicating, commanding and cajoling the dog to achieve certain behaviour. Different work requires different human attributes. We discuss educating for diversity but less about various human qualities. These are arranged into mental categories for representing on a checklist or test for comparing, tracking and ranking students, dismissing attributes that are important to life. However, for someone looking to make a living, the important thing may not be assessment scores but whether, as in the above example, they are careful or communicative people. You only have to review the huge increase in drugs, administered to reduce activity and enable people to sit for hours at a desk or computer, to understand there is now a '*one-size-fits-all*' norm that pervades society and denies the diversity amongst us all.

If different human types are attracted to various jobs, the opposite also occurs. The work someone does influences and shapes them in both intellectual and moral ways. For example, fixing things is different from building them from scratch. The medic and mechanic deal with failure and fixing, but the architect and builder create and construct. Diagnosing and fixing things, which are complex, variable and not fully comprehensible, require a mental and moral disposition. Getting it right demands that you concentrate on a conversation rather than be creative in a construction (Cannon, 1976). Cannon focused on the importance of cultivating *attentiveness*. Things need fixing and looking after no less than creating. Fixing is an active process, depending on knowledge, pattern perception and basic causes, together with the ethics of involvement. Clues and causes are sought only if a person is committed. Intellect and moral virtues are entwined. Psychologists define *meta-cognition,* which is the action of stepping back and reflecting on your own thinking, to consider whether understanding of a problem is adequate. This capacity is partly moral and not measured by intelligence tests. One has to be constantly aware of the possibility of being mistaken, which is an ethical virtue (*worthy characteristic*). Iris Murdoch (2001) pointed out that to respond to things justly one must perceive them clearly. Such *objectivism* is a virtue that demands looking beyond oneself with honesty, humility and imagination.

Without opportunities to regularly exercise judgement, attentiveness will atrophy. If you do not use it, you lose it. This questions whether degraded work results in not just a dumbing down but a diminishing morality. Recruiter surveys (2018) complain about lawless, lazy, laid-back employees, compared to the efficient, reliable, active ones of years ago! There seems to be a vicious circle in which degraded work plays a role in shaping us into careless, disengaged workers. The quality of being wide awake, clear-sighted and able to understand the whole situation cannot be taken for granted, and depends on others keeping us up to scratch. Jackall (1989, p. 11) points out how problematic this is with those who manage our lives '*using empty or abstract language to cover problems*', in order to obscure reality. In such a moral maze, it is common for them to remove themselves from the details of a production process, which facilitates non-accountability and self-estrangement.

Such issues are important to understand, and the process begins in school. Murray (2008) reports that 90% of students are encouraged to go on to college with little interest in their personal qualities, which might suggest a different pathway. Some bright people are ill-suited to both higher education and the work that follows from a degree, as demonstrated by our example of the chap with a PhD in genetics who is now working as a plumber. I also have a friend with a first-class degree in chemistry, from a top British university, who is now a bus driver, as he could not stand the suppression of intelligence in the corporate world. Pushing everyone into college can create perversities in the job market. Collins (2002), quoted by Weyrich (2006, p. 34), describes the cycle of credential inflation that '*could go on endlessly… until babysitters are required to hold advanced degrees in child-care*'. If corporate knowledge work is intellectually undemanding, then academic achievement is an ineffective way to recruit. Brown & Scase (1994, p. 52) quote a recruiter: '*We find no correlation at all between your degree result and how well you get on in this company.*' The irrelevance of what you actually learn in school for job performance is hard to equate with a common belief in meritocracy and the necessity to be academically well qualified.

Such views are known as '*human capital theory*', where '*schools help society get the skills that it needs while they help individuals get the social positions that they deserve*' (Labaree, 1997, p. 27). This is based on an idea of education that was prevalent last century. After the Second World War, life was viewed as becoming more complex, and it was assumed that graduates were needed with superior knowledge and skills. However, there has been scant evidence to show that they were better at their jobs. Berg (1970) found an inverse correlation between educational achievement and job performance. Education has been regarded as instrumental and good for society, but this tends to have a corrupting effect on genuine learning. Labaree (1997, p. 33) says: '*Formal characteristics of schooling such as grades, credits and degrees come to assume greater weight than substantive characteristics, as pursuing these badges of merit becomes more important than actually learning anything along the way.*' Students focus on grades to achieve admission to their chosen courses. What matters is how they rank amongst peers rather than their specific learning achievements.

Collins (2002) suggested that higher education rewards displays of middle-class discipline. Passing exams, meeting course deadlines and dedicated study show a willingness to conform to requirements, with qualities needed to develop competence in a bureaucracy. However, present needs for workplace *flexibility* put different demands on people. Personal qualities, supporting communication, collaboration and cooperation, are more important than certificates, which do not always prove the ability to apply knowledge judiciously. Job recruiters look for psychological and social aptitudes that are not easily defined. Nowadays, employers want evidence of personal as well as academic competencies. The ability to communicate well, establish positive relationships and work as a member of a team is considered vital, as a basis for mutual support. This enables friendship and solidarity amongst different ranks, so helping them to implement clear standards. Countries like Japan are well attuned to such requirements, fostering communication, cooperation and collaboration right through school and college, with huge benefits for workplace performance (Sage *et al.*, 2000–10).

DIFFERENT KINDS OF KNOWLEDGE AND UNDERSTANDING

Recently, there was a house fire in the road where I live, and a friend who was a firefighter was heading the team to save the building. He mentioned that they exited the building seconds before it collapsed, which led us to talk about *intuition* and the possibility of a '*sixth sense*'. Kendall (2011) reports that top moguls are cultivating their *intuitive sense* through meditation, crediting clarity and best ideas to this investment.

We have previously discussed that our current education is based on the *importance of 'knowing that' as opposed to 'knowing how'*. This equates *to universal knowledge* versus that resulting from real, *individual experience*. Universal knowledge is transmitted through speech or writing, and this kind has prestige. Practical know-how, in contrast, arises from experience, and cannot be downloaded but only lived. The things we know and understand best are those that we engage with practically. Heidegger (1996) noted that we come to know a hammer not by looking at it but holding and using it. His insights into everyday thinking illuminate the kind of expert knowledge that is *situated*, like the firefighters' *intuition*. If thinking is bound up with self-communication (*inner language*) and action, then understanding the world intellectually depends on us doing things in it. Blinder (2006) and Levey (2006) have shown the likely consequences of jobs that are inherently situated and cannot be reduced to rule-following. These are dismissed by educators, but are most likely to bring satisfaction and rewards because they engage the brain and imagination.

IMPLICATIONS FOR TEACHERS

The 2008 banking crisis illustrates how any job that is depersonalised and answers to forces remote from the workplace is vulnerable to degradation, requiring the person who does the work to suppress better judgement. The appeal of trades is that they resist a tendency to remote control, because of being situated in a particular context. In such jobs, face-to-face interactions are still normal, rather than remote contacts by email, Twitter and Facebook. Each person is responsible for their work and clear standards provide cohesion for colleagues, in contrast to the more manipulative relations of an office team.

People involved in practical work experience failure regularly. Those that create, construct, fix and mend have to contend with complex situations where one thing sets off another. This was the case recently with our dishwasher, when the springs went, damaging hinges and letting water into the electronic panel. It required three engineer visits, as the problems were complex. However, the people who make decisions that affect us often have little sense of their fallibility, which results in lack of caution. When I first was employed in a hospital rehabilitation centre, I was told by the director that no one was assigned to work there who had not demonstrated coping with major difficulty and disappointment. People had to be aware of what happens when things go badly wrong, as they do in a context where minds and bodies are disrupted and emotions distressed.

Murray (2008) argues that experience of failure has been edited out of the educational process, especially for able students. With grade inflation and soft curricula, students can complete studies without ever being seriously wrong. Such education merges with effects of our present material culture and insulates us from confrontations with hard reality. It is important, therefore, for able pupils to have practical experiences and have their egos bashed before eventually running organisations. Failing experiences teach us to pick ourselves up, so that when achieving we feel pride and satisfaction. When I worked in Japan, I found that teachers saw failure as the best way to learn, with students encouraged to verbally explain how a task was done rather than consider it right or wrong. This philosophy is in stark contrast to British practice.

For the sake of student futures, teachers must be regularly reminded of the different expectations of workplaces when compared to those of schools and colleges. They have been forced to concentrate on products of learning when the process and personal development of students is what will be important for future earning power. Some cultures have achieved this balance and find their citizens more employable as a result. The world has changed, and education must take this on board in helping student awareness of what is expected of them and the competencies needed to be effective employees. Blinder (2006) suggests that the future labour market will be between '*personal and impersonal services*'. The former requires face-to-face contact or a link to a specific site (*doctor or dental surgery*). The latter refers to activities farmed out abroad, like customer service call centres. Levey (2006) discusses the issue not in terms of whether a service can be delivered electronically but if it is rules-based. He points out that until recently one could make a good living from following instructions, like preparing tax returns for businesspeople. Now this work is attacked on two fronts – some goes to offshore accountants and some is done by software, like TurboTax. The result is downward pressure on wages for jobs based on rules. This urges us to consider the human dimension of work and in what circumstances it remains indispensable and why. It is about knowing what to do when rules are insufficient or non-existent, requiring imagination and intuition. Teachers have a vital role in helping students to orient themselves to different workplace demands, making them aware of the possibilities and how they might prepare for these. A finding of the 2009–19 Career Builder surveys is particularly important. This was the value placed on '*poor communication skills*', which employers cited as a deal-breaker. The high-achieving nations understand this. Cuba, high in UNESCO tables, cited in our first example, puts priority on this in education and, like Japan, ranks communication and relationships as top of their agenda. Understanding what the job market requires is vital for providing education that fits people for work in the new industrial age of artificial intelligence (AI).

REVIEW

AI is today's big buzzword in the work world. In the networked digital era, amid the proliferation of smart devices and surge of Big Data, smart machines that think intuitively and make sense of vast information are the new battleground. They disrupt companies and build rivalries amongst tech titans – Tesla's Elon Musk and Facebook's Mark Zuckerberg. Digital giants, such as Google and Amazon, now spar with AI-based virtual assistants. Driverless cars and robotic assembly plants are disrupting the automobile industry and the manufacturing sector. This scenario is shaking up the work world. The coming job losses are only part of the story. Automation (*driven by AI*) threatens jobs world-wide, according to the World Bank. However, AI proliferation means different work roles will evolve, but robot automation ensures that everything that can be programmed will be. How people will prepare and seek work and employers hire them will change drastically. Now 70% of employers use social media to screen candidates for prospective jobs. This means that people need to think before they act and not do anything that prejudices their reputations. Schools are directed towards a prescribed learning agenda for students, but must be creative in developing personal competencies, which the new age demands for flexible working. This can be done if they teach in participatory ways to involve students actively, using real experience to achieve what the curriculum prescribes. This is so important in our increasingly competitive world.

MAIN POINTS

Teachers need to:

1. Give students chances to engage in practical tasks to learn what is involved in producing something from start to finish for reflecting on competencies needed to see a task through, like self-reliance, communication with oneself and others, relationships, persistence, commitment, resilience and integrity.
2. Help students to become aware of their personal, unique qualities and to base their future career decisions primarily on these dispositions.
3. Enable students to view failure as a positive experience that increases rather than decreases learning.

4. Develop knowledge of the workplace by contact with career specialists, courses or other disciplines and bringing this information and experience into schools.

REFERENCES

Abrahams, D. (2013) PEEP Presentation: Sofia, Bulgaria, 16 April

Arendt, H. (1958) *The Human Condition*. Chicago: Chicago University Press

Aristotle. *On Generation and Corruption*, 316a 5–9

Berg, I. (1970) *Education and Jobs: The Great Training Robbery*. New York: Praeger Publishers

Blinder, A. (2006) *Offshoring: The Next Industrial Revolution? Foreign Affairs* (March/April 2006)

Braverman, H. (1974) *Labor and Monopoly Capital: The Degradation of Work in the Twentieth Century*. New York: Monthly Review Press, pp. 74–76

Brown, P. and Scase, R. (1994) *Higher Education and Corporate Realities: Class, Culture and the Decline of Graduate Careers*. London: UCL Press, p. 138 onwards

Cannan, E. (1976) (ed.) *The Wealth of Nations*. Chicago: University of Chicago Press. Bk. 1, Ch. X, Pt. 11, pp.142–43

CareerBuilder Survey (2009) Press release 19 August, 2009, www.careerbuilder.com

Clifton, J. (2017) *The World's Broken Workplace. A Review of the 1917 Gallup Poll*. news.gallup.com/.../world-broken-workplace.aspx (accessed 6 August, 2019)

Crawford, M. (2009) *The Case for Working with your Hands*. London: Penguin

Garson, B. (1989) *The Electronic Sweatshop: How Computers are Transforming the office of the Future into the Factory of the Past*. New York: Penguin, p. 120–121. Workplace Gallup: www.gallup.com/workplace/2590p76/leadership-rules (accessed 6 August, 2019)

Grafstein, D. & Nelson, B. (2019) *3 Leadership Rules that Separate Good from Best*.

Heidegger, M. (1996) *Being and Time*. Translated by J. Stambaugh, Albany, NY: SUNY press, I.iii, 15, p. 63

Kendall, P. (2011) *Power Yoga: Meet the Mediating Moguls*. London: The Daily Telegraph SEVEN, 6 November

Jackall, R. (1989) *Moral Mazes: The World of Corporate Managers*. Oxford: Oxford University Press

Kojève, A. (1989) *Introduction to the Reading of Hegel: Lectures of the Phenomenology of Spirit*. Ithaca, NY: Cornell University Press

Labaree, D. (1997) *How to Succeed in School without really Learning*. New Haven, CT: Yale University Press

Levey, F. (2006) *Education and Inequality in the Creative Age*. Cato Unbound, June 9, 2006. Available at www.cato-unbound. org/2006/06/09/frank-levey

Murdoch, I. (2001) *The Sovereignty of Good*. London: Routledge Classics, p. 21

Murray, C. (2008) *Real Education*. New York: Random House, p. 103

Pells, R. (2017) *Growing Numbers of Students Regard Universities as 'Poor Value for Money'*. The Independent. www.independent. co.uk/student/news/higher-education-students-university-poor-value-money-dissatisfaction-survey-poll-a7779161.html

Plato: Gorgias, 465a

Recruiter Survey Results (2018) *Recruiting Headlines*. recruitingheadlines.com/2018-recruiting-survey

Rose, M. (2005) *The Mind at Work*. New York: Penguin

Sage, R., Rogers, J. and Cwenar, S. (2006) *Why do Japanese Children Outperform British Ones?* Project 2: The Dialogue, Innovation, Achievement and Learning Project. Leicester: University of Leicester

Turner, J. (2013) *Schools Like Mine Have Become a Dirty Word*. London: *The Week*, 23 February

Weyrich, N. (2006) *The Pennsylvania Gazette*. March/April 2006 in Collins, R. (2002) *The Chronicle of Higher Education*

7

COMMUNICATION
COMPETENCIES

PREVIEW

In 1890, in Volume 1 of his scientific papers, James Clerk Maxwell, physicist and mathematician, wrote, 'Happy is the man who can recognise in the work of today a connected portion of the work of life and an embodiment of the world of eternity' (Niven, 2012, p. 12). Education typifies the past, present and future and is crucial in assisting the old and new generations to acquire the abilities to survive. Today, changes in living are faster than ever before, due to the advent of intelligent machines that are taking over mundane tasks and creating opportunities for new human roles. AI is being used to recruit staff, using verbal and non-verbal measures in mobile phone interviews. However, the educational curriculum has been narrowed over the last 30 years to elevate scores on national standard tests. These dominate teaching, leaving little room for anything else, although the independent education sector is moving to change this, by encouraging a broader curriculum that focuses on transferable abilities. In professional training, courses have increased preparation in subject matter and minimised personal, professional development. A political agenda has trumped best practice and led to a generation that is neither fully ready for life nor work challenges. The chapter summarises this discussion and poses solutions that are rooted in a more focused attention on communicative competencies to ensure improved confidence, coping, co-operative and collaborative attributes for the 21st-century citizen.

INTRODUCTION

This book has been written because an update was suggested for an earlier series on communication issues (University of Leicester, 2000–7). Research showed that many children entered school in this area with communication and thinking at least two years below their chronological age, which increased at senior school entry, because this problem was not identified or dealt with during learning. Research in 10 Leicester schools, to investigate why students failed to make effective academic progress, indicated that communication with adults was the greatest learning problem, as teachers spoke in unfamiliar ways (Sage, 2000).

Communication failure has grave consequences. The INTERMAR project, discussed in earlier chapters, showed that this was a reason for disasters on land, at sea and in the air, with courses in intercultural communication seen as essential to mitigate problems. The 2016 CRICO report into deaths at Harvard hospitals demonstrated that 1/3 were due to communication failures, which could have been avoided if more attention had been paid to the process. Certainly, there is misdiagnosis because professionals are not trained to monitor communication issues. A review of children referred over 10 years by schools for unacceptable behaviour to a Paediatric Assessment Centre showed that all had higher-level language deficiencies affecting their ability to assemble quantities of talk and text for understanding and expressing ideas. These students would rather be thought 'bad' than 'dim' (Sage, 2000).

There is a story about the Hyatt Regency Hotel, Kansas, with amazing views from the fourth floor, overlooking a splendid lobby. Highlights were the walking paths – suspended in the air. One evening, in 1981, the walkways came falling down. Over 100 people died and over 200 were injured. The collapse was, until that of the World Trade Center, in 2001, the largest structural failure in American history. The mistake was not a conspiracy or planned act – like the terrorist one of 9/11 – but the consequence of a common mistake: ineffective communication. The company proposing the design thought it was solid. The contractor producing it pointed out flaws and solutions, which were dismissed and ignored.

Does this sound familiar? How many times have you had to implement a new scheme and found things needing change and tried to communicate views? Iterative design, like iterative development, only works when you legitimise feedback. This must be focused and made a priority. In this case, there was no response to feedback, and the result of ineffective communication was fatal. Many examples of communication failures are a by-product of:

- Not giving enough time and attention for feedback
- Not believing there is something to learn from feedback
- Not systematising a process for regular feedback
- Not ensuring the right experts are influencing policy and implementing practice

Failing on these fronts results in ineffective communication with undesirable consequences – which can be hurt feelings, limited performance and even death. The following discussion reiterates messages of former chapters and focuses on improving communication. In Japan, a Leicester team researching development of 21st-century citizens (2000–10) found that teachers coached students constantly in all lessons to improve communication. They pointed out the different styles required for various purposes and the importance of matching verbal and non-verbal messages. We have all witnessed responses to a child's picture or model – *'That's nice!'* – showing distain in the tone of voice, which negates the words.

Are You Waiting for a Disaster to Happen?

Working with schools to improve academic performances, they often ask what strategies to use. Is there something to assist distributed teams? Can they plan to maintain awareness of processes requiring priority? Unfortunately, strategies alone do not solve problems, and may make matters worse. Human knowledge and understanding are required. How does one make sure disaster is not around the corner? Three ideas help:

Meet more to meet less. It seems counter-intuitive, but short, daily 'pulse' meetings with colleagues reduce the need for more. This practice is common in Japanese education (*Hansei – reflection sessions – happen at the beginning and end of the day for **all** staff*).

Make assumptions explicit. Assumptions are made about who knows what in organisations. Challenge people to take what they assume others know and own information sharing, even if thinking they have this already. It is surprising who lacks knowledge.

Communication is a choice. Whether motivated by fear, unwillingness to cause conflict, etc. – not communicating honestly is a choice with costs and consequences worse than getting over fear. Reflect on the effects of bad communication:

- Lost effectiveness
- Misunderstanding
- Unnecessary conflict
- Broken relationships
- Inadequate performances

Never be afraid of overcommunication, as it seldom occurs. We only take in about 12% of what we are told (Sage, 2000, Setty, 2010).

DOUBLE RESPONSIBILITY

Communication involves two groups: senders and receivers, with each responsible for ensuring effective exchanges. If either or both parties fail there will be a breakdown. Since communication always involves at least two individuals, who bring biases into any exchange, misunderstandings can occur at either sender or receiver end. Discussing with students their responsibility for understanding subject content brings an awareness of communication failure and how to prevent it.

Knowledge assumption
What is obvious to you may not be to audiences. Skimming over key details can leave them bewildered and frustrated. It is difficult to know what knowledge listeners bring to the communication, so it is important for a speaker to give a clear, comprehensive explanation, whilst monitoring listener clues of confusion, like disruptive behaviour. Listening to and addressing questions while understanding where confusion lies is not a common ability.

Inappropriate language
Using inappropriate language is another problem. Technical jargon can cause confusion. Also, imprecise language leaves an audience with incomplete understanding or misunderstanding. Finding balance between technical language for a message and using comprehensible, stimulating vocabulary is the key.

Non-receptiveness
As a message receiver, it is vital to be receptive and understand what a speaker is saying. This is complicated by personal biases or boredom. If disagreeing with what is communicated, it may be difficult to listen with an open mind. If a subject is uninteresting, it is easy to tune out the speaker – missing key ideas. Active listening, taking notes, or repeating and paraphrasing if speaking, helps message reception.

Ego

We may feel it is more important to be right than to understand other views. Instead of listening, we spend time when another speaks thinking about our response. If a conversation degenerates into argument it is easy to use this tactic. Instead of understanding a person's ideas, we concentrate on how best to impose ours on them. Separating ego from an argument and listening to others is difficult but vital for effective communication. Some believe that being right about something is enough reason for others to pay attention to them. While satisfying an abstract principle, it seldom works. We pay more attention to those who are polite and respectful and tune out abrasive, rude people. Effective speakers treat listeners as equals, worthy of courtesy, with the attitude that the attention of others is a privilege, not an entitlement.

These aspects cause a roadblock to communication. A *language barrier* refers to not being able to speak the same language, but it applies to anything regarding verbal/non-verbal communication that prevents comprehending each other. Students must understand this fact.

What is a language barrier?

An English-speaker travelling in the Middle East and meeting people only speaking Arabic faces a language barrier. Arabs may understand gestures for things like *'drink'*, but a shared verbal language does not exist. Other barriers relate less to literal *'language'* and more to dialectal issues, slang or lack of shared vocabulary to discuss a topic. For someone lacking a business background, discussing collaboration with a chief executive is a barrier because no shared vocabulary leaves them without the ability to connect. Language barriers impede business deals, relationships, an emotional understanding or creative collaboration, causing misconceptions that can lead to conflict, frustration, offence, hurt feelings and even violence.

What are other barriers?

Although language is the common obstacle to communication, there are other limits. In *Class Talk* (Sage, 2000) these are discussed for classrooms, but here a broader approach is taken. Barriers exist if oral delivery is unclear. Communication only occurs when a listener hears and understands a message in the way it is meant to be received. Problems in oral communication include using unfamiliar words and those with ambiguous meanings. A teacher must ensure students understand word meanings. Another issue is using generalisations and stereotypes. Classroom communication should be specific to the topic and without bias. Teachers must not make premature conclusions before having all facts about a topic or situation. Finally, they need to overcome any lack of confidence and deliver the message with assertion and clarity, using a neutral accent for easy transmission of ideas.

Physical barriers also impede communication and understanding, like walled spaces or ways a person has to approach another for speaking. Difficult terrain, high security, inaccessible locations or places with inadequate acoustics are all barriers. Many times I have been in a school hall and had difficulty hearing a platform speaker! These circumstances create a psychological barrier, with the person communicating finding the physical barriers taxing, so unsure that their message will be received.

The perspectives, biases or preconceived notions with which someone enters a situation also limit communicating effectively and honestly. If interacting with a prejudice, the work to be planned and implemented may be disrupted by perceptions. Views act as barriers that make it difficult for new ideas to take shape, and may mean communication between parties is impossible. Perception is a common limitation for effective classroom communication. Learners may receive and hear the same message but interpret it differently, due to their varied experiences. A teacher must focus on both positive and negative aspects of conversations. A distorted view means emphasis is only on one aspect.

Other challenges are individual emotional barriers so as not to suffer or feel pain. This is understandable, but must be dismantled in order to become someone who communicates effectively and is open to new experiences. All organisations have cultural barriers. As our digital world increasingly means that affairs are conducted globally, with people from all cultures and walks of life, it is vital to understand different values, taboos and customs. Failing to acknowledge these prevents connection with colleagues from different backgrounds.

LACK OF ATTENTION AND AWARENESS IN COMMUNICATION

Lack of communicative awareness causes setbacks and conflicts. When people do not pay attention to language used and focus on things other than the way they come across, they fail to communicate. An attentive communicator listens and looks for cues in the listener's body language, behaviour and silence, using this feedback to monitor their actions.

What is active listening?
Active listening ensures one is also communicating by reflecting back to the person what has been said. An active listener does not form a

response while another is still talking, or get distracted by what is happening around them. It is important to understand everything that the person is saying with words as well as pauses, body language and voice intonation. One must maintain eye contact with a speaker to show focus on the message. Nodding and saying *'Mmm hmm'* or *'Yes, I see'* helps them feel understood, and allows them to communicate freely.

When clarifying something another person is saying, it helps to repeat things in one's own words. Paraphrasing what has been understood creates an opportunity for correcting information that has been misunderstood or misspoken. It ensures that speakers feel listened to, and shows an understanding of where they are coming from. It is vital to legitimise the message, even if it is your view. Communication then flows because they feel comfortable, open and honest. People use talk to empathise, cooperate, rebuke, request, inform, persuade, caress, confront, attack or abuse. Symbolic speech is uniquely human. Depending on what is to be accomplished, there are many ways of speaking that limit or enhance effectiveness.

Although *hearing* what someone is saying, are you actually *listening*? There is a difference between active and passive listening, affecting relationships and job prospects and indicating personality traits, making the former important to master for a positive impression. This is paying full attention to a speaker, building rapport and trust by hearing the message behind words, not just what one thinks is being said or wants to be heard. The active listener spends more time listening than talking. They listen and respond to a speaker, either through body language or their own words, being genuinely interested in hearing, understanding and legitimising other views for engaging in intellectual exchange. Empathy is vital to validate words and recognise feelings. An active listener is motivated and open to new ideas.

What is passive listening?
In passive listening, the receiver appears to attend to a speaker, but makes no effort to understand the message. When listening to music, while busy studying or cooking, one is passive. There is an awareness of the music, but attention is on the task. A passive listener, in a conversation or learning context, may accept and retain information heard, but does not question or challenge the message or show interest through words or body language. They avoid discussion or giving opinions and are unreceptive to new ideas – talking more than listening. While active in conversation, they are actually not paying attention to what is said. Many students are passive listeners for reasons of disinterest, lack of understanding of content and ineffective speaker presentation or performance.

Active listening benefits

An active listener makes a better impression than a passive one, interacting by commenting, asking questions and reacting appropriately. The passive listener does not react, appearing bored, disconnected or ignorant. An employer seeking someone engaged, self-motivated and open-minded is likely to hire an active rather than a passive listener. In life, active listening is important because better communication facilitates effective relationships with others. This is because active listening is two-way communication between parties.

Improving active listening

To improve listening, pay attention to the other person. Face the speaker, stand or sit straight and relax your shoulders. If talking on the phone, turn away from other distractions. Mentally repeat what the other person says to retain the information and maintain focus. Acknowledge the message with verbal and non-verbal cues, like nodding, asking questions and making comments at pauses. Develop empathy by asking yourself, '*How would I feel if I was that person?*' This makes a speaker feel heard and helps produce the best response.

Translation

Verbal communication is not a one-way street, even in the classroom or lecture hall. Someone is listening. Speech efficacy is not measured by how something gets spoken or sent, but by how closely listener understanding matches speaker intent. Most misunderstandings and breakdowns originate in a speaker's failure to translate an idea into language that is understood by a listener in the way they meant.

Word choice

Word choice is important, but not everything in effective talk. Words that are not part of a shared vocabulary between speaker and listener will either not be understood or taken to mean something else. Know the audience and adjust the language accordingly and, if teaching, explain difficult words, checking regularly for understanding. Address an audience directly, neither above nor below them. Effective verbal communication is translating accurately through appropriate word choice and intelligible syntax and grammar, which is easier in face-to-face talk rather than text messages. Once I was running a course when a medical doctor participant expressed annoyance that her emails were constantly ignored. When scanning them, the use of 'phatics' was never seen. These are elements of small talk, or 'niceties', that provide a pleasant wrapping to the message, such as: '*How are you?*' '*I hope things are going well*', etc. It was suggested that these might be inserted, and a review with her after six months found a significant increase in replies, which made the doctor feel better!

Disconnections between speakers and listeners are caused by various distortions along the communication pathway. These are not merely words, though some are confusing, like time and place, surrounding conditions, pre-existing relations, differences in belief systems, expectations and sensitivities.

MESSAGE DISTORTIONS

SENDER	
MOTIVATION	PURPOSE
BELIEF	TIME
RELEVANCE	CONTEXT
SENSITIVITY	CONSEQUENCES
CONSISTENCY	COMPETITION
ROLE	INTEREST
COST	BENEFIT
	RECEIVER

Between sender and receiver the path appears straight, but this is seldom the case. There are many ways to distort or filter messages (*delivering and receiving*). Distortions occur for both listener and receiver. Improving communication requires appreciating that it is rarely perfect or clear in and of itself. We must learn to listen better and speak more clearly and check whether a message has been delivered correctly and heard properly.

Moreover, spoken communication involves *inflections,* understood through non-verbal cues, like eye contact, voice tone and body language. Saying '*no*' to a child acting recklessly, with a big smile, will not be as effective as making direct eye contact, using a firm voice and frown – matching verbal and non-verbal messages. Communication is give-and-take. An effective communicator will constantly attend to listener reactions. Listener body language and facial expressions communicate with the speaker to let them know if the message has been received. When possible, stop talking and check by asking listeners if they understand or can tell you what they think you said. Active observation and listening are key to effective verbal communication.

Teachers have a difficult job trying to communicate effectively in classes that are growing in size and may contain students from varied backgrounds. Common issues affecting effective communication are listening, perception, oral and cultural barriers. Learning to recognise and overcome these is vital for successful performances. If finding that people are not responding to words, it may be due to delivery. Effective communication is the cornerstone of professional or interpersonal relationships. The meaning of a message is about interpretation, not intent. If wanting to master arts of successful interaction, strategies can help.

Verbal abilities

To analyse communication, start with verbal abilities. A message can be through a presentation, speech or conversation, with style having significant effect on how well listeners understand. It can also determine whether an audience believes what is said. This consists of voice tone, accent, gestures, manner, appearance and bearing, as well as the words used. To select a way to relay a message, first understand the audience. If speaking to children, you would not want to use a brusque, loud voice with a strong accent and difficult words in a formal style. Regardless of the message, it will be dismissed before it has even been heard.

Non-verbal abilities

Vital elements of effective communication are non-verbal competencies. These cues include eye contact, body language, facial expressions, gestures, manner, proximity and other outward signs like sweating, yawning, shaking, general appearance and posture (*crossed arms, standing/sitting, relaxed/tense*). Also included are the emotions of the sender and receiver (*e.g. yelling, speaking provocatively, enthusiastic*). Other connections are between people (*friends, enemies, personal/ professional/philosophical/age similarities or differences, attitudes, expectations*). To maximise non-verbal language, *stand* (*weight over balls of the feet to follow the gravity line for voice projection*) or *sit* facing the receiver and making direct eye contact. Do not fidget or make distracting movements. Utilise a familiar proximity limit. Americans are comfortable when there is at least five feet between themselves and strangers, whilst some Europeans feel natural at less than two feet. Relax and deliver the message, using appropriate communication style and body language (*e.g. standing, sitting, relaxed, tense*).

Feedback

Feedback is an important element of effective communication, but is often neglected. This is vital for composing a successful message. Allowing opportunity for an audience to provide feedback, after making a point, monitors strengths and weaknesses in the communication approach. This helps to modify a message for more effective presentation. Feedback after the message is delivered makes communication a never-ending, circular flow of ideas.

Voice dynamics

Have you listened to someone with a dull, droning voice? It bores quickly and makes it so easy to lose interest and tune out the speaker. Research has shown that a lecturer's monotonous voice tone was the biggest turn-off for students (Sage, 2017). If wanting an audience to look alive and pay attention, vary speaking using voice modulation. By changing the speed, tempo and volume of how you pronounce words, you can enliven your speech, create emphasis and captivate your audience. Listening to speeches and using exercises achieves this. With age, voice range diminishes, so it is vital to be aware of this and work to sound interesting.

What is voice modulation?

Voice modulation assists communication, as when raising the pitch to signal a query. It is used to question, express sarcasm and emphasise words. Without voice modulation, speaking would be done in a single pitch or tone, and it would be impossible to express doubt, query or crack a joke. When in front of an audience, fear causes effortless, natural speaking rhythms to disappear, and the voice can become strained, and words are often stumbled over, but the voice can be trained with exercises and practice.

Using voice modulation to improve speaking

The voice does more than combine sounds for words. It conveys mood, emotion and perspective. It alters word meaning by changing pitch, intonation, volume and tempo. Listeners use sound to interpret messages, and are sensitive to how voice tone affects what they hear. *'Excuse me'* can be said in different ways. A friendly tone expresses concern to gain attention of someone dropping a wallet. A louder voice and emphasis on the word *'me'* can show indignation for a person queue-jumping. When someone says they cannot help, disbelief is indicated in a query: *'Excuse me?'*, with a higher-pitched voice and rising tone. When seeing a shoplifter, the phrase is repeated quicker and louder to convey urgency and gain attention.

Social occasions are times to observe voice tone in action. As you mingle, you might hear, *'That's interesting'*, which can be interpreted variously. One voice sounds animated saying, *'That IS interesting'*, suggesting the person finds the topic fascinating. Another tone conveys shock at the opinion by asking, *'That's interesting?'* stressing the last word with rising tune. Sarcasm is detected in a third exchange by the monotone sound used with both words.

When texting, one relies on words and punctuation to get a point across, but, without hearing voice tone, this may be misinterpreted. Sending a text saying *'I didn't say Luke was there'* clouds information.

During a call, voice tone and words emphasised clarify intention. Someone else saying it is implied when emphasising '*I*', or stressing '*Luke*' suggests another person was present. Accentuating '*there*' indicates Luke was elsewhere. Although the words remain unchanged, the voice tone alters the meaning. With regard to texting, a university colleague collected 3,000 student emails in one term that had been misinterpreted. We then worked with students to discover *why* and *how* problems could be solved.

Sensitivity to voice becomes important when meeting people. A soft voice tells you the person is shy, while a loud, strong one intimates they are overconfident or aggressive. A new acquaintance is put at ease by speaking slowly – indicating pleasure by speaking melodically. However, voice tone is two-way, as when a partner responds in a monotone one may be boring them. Voice is malleable, and its rhythm and tone convey a different meaning to literal word interpretations. If wanting an audience to quieten, be calm or hang on every word, then speak with a firm, low tone. The audience will lean forward in anticipation. If wanting to annoy them, use a strong voice. Volume is used to convey passion and anger.

Try not to speak too quickly, as mixed audiences will have trouble following, so use a moderate pace. Slowing speech emphasises each word to sound serene or methodical. Speaking slowly is effective in lectures when students are note-taking, but vary pace and do not deliver information in one way. Use speaking exercises, so your voice produces correctly, memorising tongue twisters for this purpose. Work on pace, as when speaking informally, or if nervous, we speak faster, making words slurred and harder to comprehend. To prevent anyone asking you to repeat, speak at a slow, consistent pace. Go through the text in front of the mirror, or enlist help to tell you when you are speaking too fast.

Project your voice by pretending to throw it to the back of the room without shouting – checking your posture and aligning your weight over the balls of the feet. Good projection maintains proper speech enunciation and pace. Improper projection is an issue speakers unknowingly have, so remembering this is important for getting a point across and not having to repeat information.

LEARNING THROUGH LISTENING

Auditory learners make up approximately one third of students. Their learning style is characterised by listening, rather than being shown, as with visual learners, or from experience like haptic (*experiential*) ones. Auditory learners are often successful communicators because they listen well, but face difficulties linked to this style.

Retention
An auditory learner's strength is the ability to hear something and then retain it, so formats like lectures or discussions play to this. They will remember what was said, rather than read or seen, so instructions are best conveyed orally rather than in writing. They probably have a good memory for names.

Communication
The auditory learner relies on what is heard, not what is shown or experienced. They are likely to have a large vocabulary, enabling them to communicate effectively. They may appear as talkative and active in class, allowing them to succeed in assignments that focus on this ability, like presentations, debates and discussions, as well as learning foreign languages.

Easily distracted
An auditory learner can be distracted in classes. They may be unable to read quietly, preferring to read aloud. Attention wanders if a teacher shows a video or displays something without sound. Noises in and outside class, like others talking, easily divert attention from studies.

Class disruptions
Auditory learners like to be heard and disrupt if attention is not paid to their needs. They want to communicate but experience difficulty attending for long periods. This manifests as talking, whispering or humming. Without oral instructions, they may be unable to follow class tasks.

Devising lessons that result in all students learning suggests that a teacher considers *how* they process information. Different types of learners require varied methods. A student learning visually will do this best from stimuli like photos, films or demonstrations, whereas a haptic (*experiential*) one needs to move around and do things. Once a teacher and student have assessed together how to acquire knowledge, they can structure work accordingly.

SCAFFOLDING

Scaffolding is a technique used to build connections for learners by establishing details surrounding a unit before teaching. It allows a teacher to build a bridge from learners' current knowledge to information being taught. It is performed by a teacher modelling a task and gradually transferring this knowledge to learners so they can grasp the subject.

Engaging learners

Scaffolding engages learners actively. The teacher builds on the learner's knowledge of a topic – for instance in research in which the learner finds the solution to unanswered questions, motivating and urging them to study more.

Minimising frustration

Scaffolding minimises individual difficulty and can be used to calm those frustrated when learning with peers. Behaviour is monitored for learning frustration in class groups. This method may be disadvantageous, giving up teacher control to allow learning at a student's pace. This is time-consuming, so one may be forced to cut short time allocated for each student to accommodate everyone. This can result in frustration, with the urge to learn slowly fading. To handle learners in scaffolding lessons, teachers need training. The strategy requires a teacher to allow students to make mistakes in order to learn. Those untrained in the method are unlikely to intentionally allow pupils to make errors. Scaffolding is a common approach in Japan, as teachers never treat student responses as wrong but as chances to learn. This means that students are not afraid to articulate their ideas, as they are never treated negatively.

LEARNING THROUGH OBSERVATION

Albert Bandura (1977) developed *Observational/Social Learning Theory*, based on objective and subjective experience. Educators find this valuable, as they must reach a number of students to convey a central message. They can apply observational learning to improve outcomes, produce desired behaviour and enhance motivation and self-perception.

Eliminate distractions

Minimise the effects of competing sensory stimuli, like closing doors, to lessen sound interference. Some contexts are over-stimulating. In Japan, classrooms are plain in comparison, but often contain a beautiful flower arrangement to introduce beauty and create a positive atmosphere.

Present effective models

Modelling allows students to observe task stages (*making a calendar*) by paying close attention to the process. Communicative, competent, confident models receive more attention than those without these qualities. Students notice others who have the same characteristics. A variety of models (*teachers, students, males, females, etc.*) ensure students can identify with one or more of these.

Describe actions and consequences

Explaining actions and consequences means that ideas presented are reinforced. Telling students *what* is happening, *why* and *how* helps retention. Through verbal explanation, they are helped to code information, thereby increasing the possibility that they will reproduce targets.

Realistic goals

Students are unlikely to be able to reproduce a symphony from watching a musician, but observing helps them perform better while learning the basics for then mastering complex skills in similar ways.

Motivate students

Reward students for reaching goals. Through positive reinforcement of actions required, students are likely to perform them again. This may be through praise or recognition. A cycle of success means students will repeat actions without prompting.

Enhance confidence

Once students achieve attainable goals and receive genuine praise, they see themselves as more competent. Confidence leads them to future success. As a result of this positive cycle, students learn to self-regulate and become aware of their abilities.

It is a common misconception that grading and assessment are the same. While school assessment assigns grades, there is more to it than this for teachers and learners. By assessing *what* a student knows and *how* they learn and compare to peers, the teacher and student can set appropriate learning goals. The priority is that assessment helps them master steps successfully, in the way they understand best. Unfortunately, standard testing ignores the fact that students do not learn and respond in the same way and acquire knowledge and skills at different rates. Although a competitive system may encourage progress, it discourages those unable to learn what is expected.

What a student knows

Assessing knowledge and understanding is not simple. Students must *express* what they know for teacher *evaluation*. Whether verbally, through writing or other methods, a student demonstrates knowledge. Using varied assessment methods reaches all types of learners.

How a student compares

Standard testing compares students across ages and year groups. It has come under fire for negative impacts on minority and lower socioeconomic students. Many object to the way in which resulting data is used. Assessments should improve teaching and learning, rather than penalise low-scoring students and schools. Cheating results from league table comparisons.

Improvement

Learning is assessed incrementally over time to evaluate teaching efficacy. Pre- and post-project assessments determine how much knowledge a student possessed going into the learning experience and what was gained. A student may fail to achieve a set standard for many reasons that need evaluation but still make learning strides and become successful.

Goals

Setting realistic, individualised goals results from assessment. While peer comparison establishes the appropriate grade level and academic placement, it is assessment of improvement that demonstrates learning capacity. Carefully designed assessments are vital in determining future strategies for teaching and learning. Beginning in primary and continuing through to senior levels, students undergo many standard tests. Cognitive and linguistic abilities are the basic learning processes, like creativity, logic and reasoning through language. Testing these aspects measures the ability to learn, predicting academic success. Achievement tests focus on specific subjects. Both formats provide insight, highlighting areas for improvement, but, as snap judgements, should be used cautiously. Assessing progress from observing daily activities brings a more valid, reliable evaluation that takes account of more personal attributes.

Cognitive and linguistic testing

Cognition refers to mental processes, including memory, problem-solving, decision-making, learning and attention – underpinned by linguistic ability to understand and express. Lack of attention to language levels means cognitive competencies are often not adequately strengthened to process, analyse and retain information. Common cognitive tests include the Stanford Binet Scale, the Otis-Lennon School Ability Test, the Cognitive Abilities Test and the Illinois Test of Psycholinguistic Abilities (ITPA), to name but a few, but do not test narrative language levels for learning. A narrative test, like the Sage Assessment of Language and Thinking (SALT) assesses learning processes and potential in a more accurate way.

Achievement testing

An achievement test measures student knowledge and academic abilities in a standard way, typically through multiple-choice questions. Most tests focus on specific subjects, like science, mathematics and

English, etc. Common achievement tests, used clinically, include Iowa Tests of Basic Abilities, Stanford Achievement Tests and Metropolitan Achievement Tests.

Assisting learning difficulties
Unlike achievement tests, cognitive and language tests highlight learning problems and provide possible solutions. Instead of measuring student academic levels, they identify and pinpoint strengths and weaknesses. By examining underlying abilities, these tests can determine specific learning difficulties. This helps to understand why a difficulty exists, and provides answers for how to overcome it. However, many testing regimes need to be conducted by people trained and experienced in interpreting such procedures.

BROADER ASSESSMENTS

Achievement tests are used for more than measuring learning ability. They give insight into teacher effectiveness, the quality of an academic programme and information on progress. This can be used for groupings and placement, funding and teacher assessments.

Considerations
While the many cognitive, language and achievement tests measure *aspects* of student performance, they do not assess all knowledge and ability indicators. They fail to measure motivation, creativity, cooperation or attention as vital aspects of academic performance. Sport psychologists, as experts in motivation and performance, advise teachers on these attributes in some countries. Most systems use observations, tests and class participation to establish broader assessments. It is a misconception that grading and assessment are one and the same. While school assessment involves assigning grades, it is more than that for teachers and learners. By assessing what a student knows and can apply, how they learn and compare to peers, the appropriate learning goals can be set.

THE PRESENT SITUATION

The 2016 British Chambers of Commerce (BCC) survey of 3,000 firms found 90% thought school leavers were not ready for employment, and more than half said it was the same with graduates. The top five skills and attributes that employers look for in young recruits are *people skills, taking responsibility, being honest and ethical, problem-solving and team-working*, which are all rooted in *effective communication*. This is why it is important to focus on how this process can be improved. Inadequate communication is a major issue, which impacts on social interaction, cooperation and the ability to think through a task and understand the consequence of actions. The BCC Director General, John Longworth, said many businesses took the view that hiring a young person was a risky move. He said in the report:

'Businesspeople tend to favour more skilled and experienced applicants – and while they do sympathise, their primary function is to run a business, which means making business decisions. Firms need young people that are resilient, good communicators and understand how to work as part of a team. We believe that successive governments have failed our young people by not properly equipping them for their future careers… Government and educational institutions must be more focused on equipping young people for the workplace, and in turn businesses must be more willing to give them a chance.'

Ya Saffie Ceesay, a third-year Birmingham student, was quoted: *'This employability gap is really worrying for graduates. I hadn't realised that I will need to invest in my own career development; I thought employers would take on this responsibility. Leaving university this summer and starting my job hunt, it's worrying to hear that I might not have the kinds of skills that employers are looking for, even after years of studying.'*
In fact, 90% of employers expect graduates to be responsible for their career development. The Office for National Statistics (ONS) data shows that 49% of graduates end up taking non-degree-level roles as they struggle to match competencies with those expected by employers. This has an impact on business productivity and graduate prospects. To help young people close this employability gap, the Chartered Management Institute (CMI) has launched the Future Leaders Network, to help access development services and find a foot on the career ladder. Communication methods are core to this strategy, based on research suggesting effectiveness owes 7% to words, 38% to voice and 55% to non-verbal cues. Emphasis is on not *what* you say but *how* (Mehrabian, 1971).
A dull message delivered by a charismatic person, filled with energy and enthusiasm, will be accepted as brilliant. An excellent message delivered by someone not interested in the topic fails to engage the enthusiasm of its intended audience. A classic example of effective verbal communication is Martin Luther King's *'I have a dream'* speech (1963). Why was it great? It was filled with powerful visual images provoking strong emotions, delivered with passion by someone capturing the dreams of an entire race: *'I have a dream that one day*

down in Alabama little black boys and black girls will be able to join hands with little white boys and white girls as sisters and brothers…'
Over time, the speech has transcended its original message to be one of hope for all, regardless of race. The elements are summarised below:

Speaking:

- Body language
- Voice quality
- Intention
- Manner: directness, sincerity
- Dress and clothing (style, colour, appropriateness for situation)
- Visual aids, animation
- Eye contact
- Emotional content, energy, strength
- Self-concept
- Concept of others
- Listening to hear the underlying message
- Speaking from the heart
- Energy
- Setting, time, place, timing
- How the messenger holds the message
- Sensitivity
- Rhythm and pacing
- Attitude and confidence
- Rapport
- Agenda
- Purpose – knowing what to communicate
- Clarity
- Silence, focus
- Responsive to feedback

Listening:

- Attentiveness to speaker
- Eye contact
- Intention be fully awake and aware
- Openness: to other person and yourself
- Paying attention
- Listening to yourself
- Feedback
- Body language
- Change in pattern
- Expectations about person speaking – message and agenda

Try incorporating these elements into a **one-minute speech**. Tips are in the appendix to this chapter!

Carl Rogers, in *On Becoming a Person* (1961), notes that all psychotherapy deals with communication failure. The barrier to interpersonal communication is a tendency to judge, evaluate and disapprove of the words of others. Real communication occurs when listening with understanding – to see the idea and attitude from the other person's point of view, sense how it feels to them and achieve their frame of reference in regard to the topic. Techniques for achieving such an understanding include *perception-checking*. Try this exercise:

Person 1. Start talking about any subject for six sentences.

Person 2. When the first person stops talking, repeat back what you *thought* you heard, with a phrase like: *I want to be sure I understand what you are saying*.

Reverse roles: Person 2 speaks for six sentences, then Person 1 perception-checks.

Practising such techniques gives respect to the person speaking and shows one understands what they are saying. If misunderstanding, both can work to clarify the message.

Don Gabor, in *Speaking Your Mind in 101 Difficult Situations* (1994), gives talk tips:

- **T** = Think before you speak
- **A** = Apologise quickly when making a blunder
- **C** = Converse not compete
- **T** = Time comments
- **F** = Focus on behaviour – not personality
- **U** = Uncover hidden feelings
- **L** = Listen for feedback

DOS AND DON'TS FOR T-A-C-T-F-U-L STRATEGIES

DO be direct, courteous and calm
DON'T be rude and pushy
DO spare others your unsolicited advice
DON'T be patronising, superior or sarcastic
DO acknowledge that what works for you may not work for others
DON'T make personal attacks or insinuations
DO say main points first, then offer details if necessary
DON'T expect others to follow your advice or always agree with you
DO listen for hidden feelings
DON'T suggest changes that a person cannot easily make

Experts tell us that public speaking ranks highest on lists of situations people fear most (*followed by death*). Overcoming this requires education and **practice, practice, practice!** A disparity in methods of delivering messages is why it is so difficult not only to speak but to write something that is clearly understood by large audiences – **only 7% effectiveness** is achieved by **words alone!** That is why good visual presentation (*graphics, colour, balanced design*) adds to a written message. These additional clues help compensate for non-verbal aspects of a written message by triggering emotions. Without such non-verbal clues, the Internet would fail miserably as an effective communication tool.

What are the differences between the following graphics and words themselves?

Click Click *click* **CLICK** CLICK

Which attracts your attention? Consider this example when developing material for presentations.

Communication for leading others

1. Imagine that you are away for six months and unable to communicate.
2. Leave a one-page guide (*beliefs, philosophy, values*) on how to carry out your role.
3. Share it with colleagues.
4. Ask if they understand it and can adhere to the values conveyed.
5. Review and revise as necessary.

This task measures clear communication effectiveness. It encourages a concise document that captures your philosophy. It shows how to translate feelings into communication that others can use, understand and learn from. *Leadership beliefs* in a training module are below. Compose your own. It may lead to surprising results with colleagues.

- Be honest, reliable and consistent
- Love yourself
- Trust instincts
- Respect others
- Be confident and communicative
- Keep smiling in spite of difficulties
- Share with others to form a community
- Enjoy work

- Keep learning
- Accept responsibility for actions
- Leave the world better than you found it
- Ask *why* and *why not*
- Consider problems as challenges
- See every day as an opportunity
- Be humble and grateful

REVIEW

All children should enter school with the communicative levels to be physically, mentally and emotionally independent. Research indicates this is not always the case, with widespread deficiencies also evident when finishing formal education and entering the workplace. The Children's Society's annual *Good Childhood Report* (2019) records a happiness level, using a scale of 1 to 10, by asking 2,400 respondents to rate their satisfaction with life. The mean is 7.89, showing a downturn since social media has exploded. It cites issues of appearance, social media, school pressures and declining strong friendships as driving factors. A third of 10-to-17-year-olds worry about having enough money to live and a quarter are concerned about their ability to get jobs. They are continually anxious about crime, the problems of the planet and issues of online abuse. Mark Russell, Chief Executive of the society, said at the launch of the report: '*Modern childhood is a happy and carefree time… yet for many it is not… Today's young people are becoming progressively unhappy with their friendships, one of the fundamental building blocks of well-being.*' Technical methods now replace talk as the favoured way to communicate. Screen time is making children less creative and imaginative with insufficient time for their own ideas, pretend play and interaction with others (Bouffard, 2017).

Natalie Pearson's (2019) study of 4,000 British 14-year-olds shows that only 9.7% manage the three international daily recommendations for 5-to-17-year-olds, with less time for live friendships, as well as having weight issues and depression. The guidelines include **one** hour of exercise, no more than **two** hours of screen viewing and **eight** hours of night sleep. This suggests that many youngsters will not have attained the competencies for employment. In an article: 'How Did Your Interview with the Robot Go?', Hymas (2019) discusses how artificial intelligence (AI) is now used to recruit staff. Unilever has 250,000 applicants for 800 posts, and AI allows applications to be sifted quicker, reducing recruiting time by 75% and saving £1 million. Candidates are filmed over their mobile phones, answering eight questions about themselves and the job. HireVue, the company behind this technology, says it removes human biases, but Anna Cox, a Professor of Human-Computer Interaction, suggests it rules out some who could do the job. It focuses on verbal and non-verbal responses, analysing 25,000 bits of facial and linguistic information. If not a competent communicator, you are disadvantaged, as language is the most important trait analysed by the algorithm, including diction, speed and tone. Candidates are ranked from 1 to 100 as good, moderate or weak, and invited for interview based on this analysis. This development is a sign of the times and confirms employer priorities: that they are looking for effective communicators in the people they appoint to job roles.

Children are not allowed to be bored and expect to be always entertained, so lack chances to be creative with their minds and bodies. Producing imaginary friends and inventing storylines are becoming things of the past, say experts. Therefore, youngsters are not achieving life skills – like face-to-face talking for sharing ideas and gaining balanced perspectives. Education should be more three-dimensional, concentrating equally on knowledge, its practical application and personal competencies to achieve success. At the end of schooling, students must demonstrate competency in subject knowledge for problem-solving, as well as abilities to relate to others and be responsible citizens. England is a conservative nation, with students submissive to curriculum demands, and often turbulent because they do not have the communicative levels to follow requirements effectively. Governments may be sincere in their mandates, but this often means they constrain practice and professional training, resulting in education that is dictated by league tables rather than learner needs. We should not be led into other ways of thinking that are unhelpful. It is important to challenge with proper evidence to support action. This book provides some facts, which must be cautiously evaluated as merely a snapshot of a situation, but should encourage us to reflect and review ideas. Changes in practice must involve all stakeholders for consistent approaches. Passion, persistence and patience is required. Practitioner doctors at the University of Buckingham are spearheading interesting developments to remedy such problems and narrow the gap between education goals and workplace demands.

MAIN POINTS

- There is a mismatch between education and work requirements
- Research is often ignored or dismissed if it does not fit a political agenda
- The rise in physical, mental and emotional health issues indicates a society in crisis

- There is hope if personal as well as academic human development has focus
- Stakeholders need more knowledge and strategies to assist progress

REFERENCES

Bandura, A. (1977) *Social Learning Theory*. New York: Prentice Hall

Bouffard, S. (2017) *The Most Important Year*. London: Penguin Books, Random Press

British Chamber of Commerce Employer Survey. (2016) www. britishchambers.org.uk

Children's Society Good Childhood Report. (2019) www.childrensociety. org.uk/good-children-report (accessed September 2019) .

CRICO Strategies National CBS Report (2016) Boston, MA: Harvard Hospitals 02215Luther King, M. Aug. 28, 1963, Lincoln Memorial, Washington, DC

Gabor, D. (1994) *Speaking Your Mind in 101 Difficult Situations*. New York: Simon & Schuster

Hymas, C. (2019) *How Did Your Interview with the Robot Go? The Daily Telegraph*, 28 September

Maxwell, J. (1890) *The Scientific Papers of James Clark Maxwell*. https://archive.org/details/scientificpapers01maxw

Mehrabian, A. (1971) *Silent Messages*. Belmont, CA: Wadsworth

Niven, W. (Ed.). (2012) *The 1890 Papers of James Clerk Maxwell*, Vol. 1 & 2. Cambridge: Cambridge University Press

Pearson, N. (2019) *Prevalence and Correlates of Meeting Sleep, Screen-Time and Physical Activity Guidelines Among UK Adolescents*. *JAMA Pediatric Journal*, August 2019

Rogers, C. (1961) *On Becoming a Person*. New York: Houghton Mifflin

Sage, R. (2000) *Class Talk*. London: Bloomsbury

Sage, R.(2017) *Paradoxes in Education*. Rotterdam: SENSE International

Setty, R. (2010) *7 Reasons Why Many Smart People Have Trouble Communicating Their Ideas*. rajeshsetty.com/2010/05/05/7-reasons-why-many-smart-people-have-trouble-communicating-their-ideas/ (accessed 25 August 2019)

APPENDIX: SHORT SPEECHES – ONE MINUTE

Despite brevity, a short speech can feel like an eternity if not prepared! Unlike longer, formal ones, they are a fun way to encourage public speaking ability. Choose a topic of interest, as enthusiasm makes it easier to deliver effectively. Speaking from the heart is the key.

Benefits of mastering impromptu speeches
A short speech is usually one with little or no preparation time. Educators use them to help students prepare for situations where they might be expected to think on their feet. From a wedding toast to briefing colleagues, the ability to speak on many subjects is a useful tool.

Impromptu topics
You may have an idea of a topic and knowledge of the audience. Prepare by picking subjects you like and rehearse them. Abstract ideas make good impromptu talks, like *love*, *peace* or *joy*. Present something/someone you love or an activity bringing peace or joy. For a reflective approach, you might choose an example of when you were not so loving, peaceful or joyful, showing evolution. Places of interest make effective topics, because if you know about them, reasons why people might want to visit can be discussed. You can explore themes like the planet, discuss goal achievement or analyse media effects. Whatever the topic, think beyond the surface for a deeper meaning or give it a humorous spin if relevant. In an exam I was asked to give an impromptu speech on '*buttons*' and talked about my grandma's button box!

Tips for speeches
There is only one chance to make a first impression, so having speeches up your sleeve can serve you well. If talking about something you love, what would it be and what would you say? Think about humorous experiences or memorable moments to share. Rehearse the speech in your mind or write it down. You can record yourself or deliver it to a friend for feedback. Practice is key.

Speech prompts
Topics that you are comfortable with and knowledgeable about make effective speeches. Speak clearly and confidently. Vary pitch and volume to emphasise points, making eye contact with the audience. Some ideas are a favourite place/object, leisure activity, memorable incident or education benefits! Giving a speech can be stressful. Reducing this improves speech quality, so be prepared, pick a topic that interests and act confidently.

Practice, practice, practice
Preparation is key to success. If there is a time limit, a timer gives a sense of delivery pace and the pressures of a ticking clock. Practice the speech many times to feel confident with the material and to review the best delivery style.

Pick an interesting topic
You display more enthusiasm and passion for an enjoyable subject. This directly affects the presentation of material and how it is received.

The worst that can happen
Speech-making is not catastrophic. What is the worst that could happen? What would you do? Things are never as bad as one thinks. We all make mistakes. The audience is rooting for you as they want to hear your ideas.

Fake it to make it
Avoid negative self-talk. If you have not banished stress, you must fake it. If you are *acting confident* an audience assumes that you are confident, and gives positive feedback to ease tension as you progress. A common first step is to admit nervousness, but this does not set a confident tone and makes the audience uncomfortable. Smiling, breathing deeply and not rushing through the speech help you appear confident. Today, you will encounter those not speaking English well. People globally use English as a common language and linguists think there are more non-native than native English speakers. You can make interactions easier on someone mastering the language by communicating respectfully and effectively.

Make no assumptions
Someone improving their English may struggle with *pronunciation*, *accent*, *verb tense*, *syntax* and much more. Although unfamiliar with complex English phrases, they may have adequate comprehension. Do not assume learner needs until knowing them better.

Be aware of the cultural context
Each time you speak, you deliver your cultural norms. The English value directness and independence, but this may not be true of other cultures. There will be differences in your expectations of social situations and how to communicate needs, so research other norms.

Be clear and direct
No one likes to be patronised, so express yourself carefully. Speak clearly, but keep your voice even and do not shout to be understood. Avoid English idioms like '*Let's wrap this up*'. Drop unnecessary words: '*Do you mind if I sit here?*' should be reduced to '*May I sit here?*' Speak slowly if worried about being misunderstood, but slowing down too much is confusing. Ask listeners to repeat what you have said to check comprehension.

Be patient and use body language

When others speak, your behaviour communicates volumes. Allow time for them to think through responses, not supplying words unless asked. Sitting or standing facing others helps to follow the message and monitor each other's non-verbal responses for adjusting input.

Lie detection

Lie detection is a trick to seek out speaker deceptions. Face-to-face situations establish a baseline with innocuous questions, then it is possible to spot signs of evasion or deceit. What happens when speaking on the phone? You do not have speaker mannerisms to monitor. There are still cues for spotting lies even when voice is all you have. Establish a baseline by chatting casually. Attend to the speech pattern, word flow and the way sentences are pronounced. Ask about the weather, sports or similar subjects. The aim is not to interrogate, but to put the subject at ease. Listen for alterations in speech patterns. A long pause before answering means probably looking for the right words, and is suggestive of deceit. Listen to word flow for changes in tempo – speaking more slowly or quickly – or using guarded speech terms. A relaxed speaker often tells the truth (*or practised the lie*), while a nervous one is more likely to be untruthful.

Look for topic changes. A liar often avoids speaking about the topic directly, attempting to shift it elsewhere. They use evasive or non-responsive answers, addressing questions without details. Expect hostility. Arguments and accusations obscure the issue, getting a liar off the hook in '*he said, she said*' terms. Remain calm and on-topic, focusing on the issue. If bluster continues, there is probably something to hide. Check other methods of creating confusion. A liar may repeat questions, rephrase answers and seek to obfuscate the topic to escape interrogation. Stay clear and direct if this is tried, giving them less room to hide. Lie detection is instinct as much as certainty, and those lying may not be malicious. Use your judgment when ascertaining if someone is not telling the truth, but refrain from making accusations without evidence.

8

FINAL THOUGHTS: THE EPILOGUE

'The single biggest problem in communication is the illusion that it has taken place.'

George Bernard Shaw

Whatever the future, experts agree that *language and culture* will become the most important unifying factors amongst people. Education will need to take more interest in linguistic and cultural issues that presently divide populations and cause tension, distrust and problems in understanding and learning effectively. Research at the University of Leicester showed how, in city schools, children all entered with language and cognition ages about two years below chronological ones (Sage, 2000). Two decades later the situation has not changed, with teachers reporting that this continues to be a major issue in formal learning.

The basic obstacle to effective communication is the problem with meaning, which is especially important in cross-cultural interactions. What you mean when you say something is not necessarily what the listener hears. Messages derive most sense from their cultural context and non-verbal signals. In cross-cultural communication, these are composed or coded in one context, sent and then received or decoded in another ethos. It helps to become a more knowledgeable communicator to engage with others successfully. This lowers stress levels, making effective decisions and working together more possible.

Intercultural communication defines the exchanges between people embracing several cultural contexts. We are all different, so it is important to recognise each other's communication differences – customs, attitudes and values.

Cultures are classified as *individualistic* or *collectivist*. When asking a question from collectivist people, such as: '*tell me about yourself*', they are likely to talk about themselves in relation to others, such as, '*I am Japanese, the daughter of farmers, with four brothers*'. Self-identity is shaped to a stronger extent by group memberships. Members of individualistic cultures recognise themselves as distinct, autonomous, self-confident and resourceful. When you ask the same questions they would be more likely to talk about their likes and dislikes, personal goals or accomplishments.

Adrian Holliday (2016) mentioned that Australia, Britain and the United States of America have individualistic cultures, while Asian countries like Afghanistan, China, Japan and Latin America have collectivistic types. Collectivists are permanently attached to their groups, whereas individualists like to change them and have weaker involvement. In countries where they have collectivistic cultures, a manager is expected to show personal interest in employees' lives. Individualistic and collectivistic cultural understanding is an important starting point for intercultural communication, which has a direct effect on outcomes and developments in organisations like educational institutions.

With increasing diversity in all institutions, everyone needs to have enough knowledge of the local philosophy as well as cross-border customs, attitudes, beliefs and values. This capability, in light of an ongoing increase of joint ventures and other transnational unions, means that language and cultural issues must have priority in education. Supporting interaction and establishing a strong relationship amongst everyone is considered the primary factor in achieving objectives. In order to accomplish this, interactions must be encouraged. Some people may have developed communication competence, but others will not have enough capability and confidence to engage effectively with others.

Therefore, educational institutions must be responsible for developing social and interactional abilities. This will not only create a stable environment to build strong relationships but improve overall performance. Offering initiation programmes for newcomers, whether staff or students, allows open discussion of policies, strategies and responsibilities of all concerned. In addition, equality and justice can be discussed to discourage racism and discrimination. When an organisation encourages everyone to debate and share ideas, it helps them to feel valued and increases their output. Also, open communication establishes trust, which eventually brings greater satisfaction and well-being.

Effective communication has been recognised as a core factor for the success and failure of any organisation's goals, policies, strategies and objectives. However, those in charge are responsible for sharing these clearly, ensuring that everyone works to the same objectives. In this case, communicative competence is primarily important. If leaders have clarity in their communication, employees will know exactly what the organisation wants. Thus, effective communication can improve productivity and understanding amongst staff and students.

Many scholars state that success, development and achieving organisation aims is about understanding the differences between cultures with respect for them (Matteucci, 2017). If everyone has enough knowledge about other cultures, they can communicate effectively to achieve goals on time. When communication is destroyed an institution cannot thrive. A heightened level of awareness avoids assumptions, expectations, biases and prejudices. This leads to the highest levels of trust, clear thinking and collaboration between people, along with individual and team creativity. Conscious communication allows each person to become aware, to assess their best attributes and influence others to do the same. We survive because of effective communication, cooperation and courage.

CONDITIONS FOR EFFECTIVE COMMUNICATION

Conditions for effective, intercultural communication are:

- **respect** for other cultures
- **awareness** of other cultures and one's own
- **knowledge** of different cultural codes
- **social competence** to bridge differences
- **connection between persons** – social and emotional

Communication is effective if the recipient interprets the message in the way the sender means. The contact has impact because the sender and recipient understand each other and so make a connection. Language is more than only knowledge of words and use of the senses. The chance for wrong communication between persons from several cultures is larger, because of the difference in standards/ norms, values and world views. These differences can be bridged if one becomes conscious of the issues. For intercultural competence and effective communication there are, according to Pinto (2000), four conditions:

Technical condition: there must be reciprocal understanding of language communicated.

- you understand each other's language – verbal and non-verbal
- you can see and/or hear each other
- a possible interpreter is available
- technical equipment (*phone, radio, television, etc.*) functions

Cognitive condition: both parties must speak at an intellectual level acceptable to both.

- you can appreciate each other's intellectual level
- the subject of conversation is familiar

Interpretative condition: the same interpretation must be made to the words used.

- you give the same meaning to the words, gestures and mimicry used
- interpreting is more difficult if the language in which you speak is not your native one

Affective condition: both know the emotional meaning of words, gestures/operations.

- you must have the same feeling about certain language actions (e.g. *politeness codes like looking at each other directly*)
- you are prepared and want to communicate

The last condition can lead to wrong communication, because one assumes when one speaks the same language the same meanings are granted to both verbal and non-verbal expressions. To serve the affective condition, it is necessary to understand or know each other's culture. For development of intercultural competence and communication with persons from other cultures, several models have been developed (Azghari, 2015).

- Three-steps method (Pinto)
- Bridge model of values and interests (Azghari)

THREE-STEPS METHOD (PINTO, 2000)

The three-steps method of David Pinto is a practical way to bridge cultural differences. It considers the double perspective and deals with possible differences to increase communication effectiveness. Learning to look from your own cultural perspective (*knowing the standards and values*) and from the viewpoints of other cultures (*learning standards and values of the other*), it is possible to understand alternative world views and prevent the following problems:

- a restricted vision: to see, experience and interpret from your own perspective;
- attributing your own standards and values to a speaker: you cannot indicate your own wishes, restrictions and borders, because it is not clear what are the other's communication codes and cultural values.

Culture has manifest and hidden aspects. The first links to age, sex, training, ethnicity, profession and social status and is directly obvious. However, there are characteristics not directly seen, like emotions, belief, feelings and perception. These hidden cultural aspects play a vital role in communication. Acting from a double perspective means misunderstandings are fewer. The three steps are:

- Step 1: Learn your own culture-based standards and values. Which rules and codes influence your thinking, actions and communication?
- Step 2: Learn the culture-based standards, values and codes of others. Separate opinions from fact in behaviour. Research what '*strange behaviour*' of others means.
- Step 3: Think how to handle in a given situation the observed differences in standards and values. Mark borders concerning adaptation and acceptance of the other. Make these clear to everyone.

BRIDGE MODEL OF VALUES AND INTERESTS (AZGHARI)

Azghari describes a number of intercultural communication models and provides criticism. The model of Pinto, according to Azghari, is a recipe on '*how to handle others*'. Azghari developed the *bridge model* to improve intercultural conversation in three phases:

Phase 1
Review which values and/or interests play a role in a certain situation in contact with the other and verify these with them. (What should I know?)

Phase 2
Recognise which values and/or interests play a role in communication with the other and make that clear to them. (How do I handle this?)

Phase 3
Decide which values and/or interests have the highest priority for you in a given situation and behave accordingly. (How do I act?)

All three phases examine successively: *values, standards/norms and interests*.

A **value** is '*what I believe*', a **standard** is '*how we want to handle each other*' and **interests** are about '*which aims we pursue?*' Standards have not been incorporated separately in the bridge model, because they generally give a concrete interpretation of a value.

Moreover, standards can change because of interests (*own interests versus public ones*). People must assess their values and interests, which may clash with others.

For example: '*Do not smoke in the house*', but if a guest visits the ban is lifted. The result of the first phase is *knowledge*, the second is *clarifying of own standards* and awareness of norms and the third is *decision-making*.

Be curious and open-minded!
Listen first, then talk and ask questions.
Listening: pay attention to verbal expression and non-verbal signals.
Talking: avoid offending customs.
Questioning: avoid suggestive questions.

Be sincere and respectful!
Try to be empathic: feel the situation and react correctly.

Be direct and unpretentious!
Words and action must agree as much as possible.

These three phases produce insights for better understanding between speakers, because the expectations from both sides have been examined with awareness of cultural customs.

LEVELS OF COMMUNICATION (SAGE, 2000)

As a teacher it is important to assess the levels of communication in a classroom and provide tasks that demonstrate all the language and cognitive levels that will be found in any group of people. These are clarified further in *Class Talk* (Sage, 2000).

Level 1 – Limited Communication
This is typified by deficiencies of language structure, form and use and paralinguistic voice dynamics, facial expressions, gestures and movements.

Level 2 – Ineffective Communication
Unconscious, unclear, assumptions, biases and prejudices made. Limited listening leads to frequent misunderstandings and relationship breakdowns.

Level 3 – Normal Communication (Informal)
Some use of conversation moves (*asks/answers mainly closed questions, contributes ideas, shows maintenance moves – nods, smiles*) but often gives inappropriate responses to open questions and irrelevant contributions are evident.

Level 4 – Normal Communication (Formal)
Some formal verbal and non-verbal communication skills seen in narratives such as tellings, retellings, instructions, reports and explanations but performance is insecure.

Level 5 – Effective Communication
Shows full range of informal and formal communication moves but not always used appropriately with confidence and rapport.

Level 6 – Skilled Communication
Advanced communication skills demonstrated such as building rapport, with informal and formal communication used, plus negotiating skills are performed appropriately with ease.

Level 7 – Aware Communication
Shows personal mastery free from emotional and irrational reactions. Communication is conscious, collaborative, considerate, thoughtful and empathetic, with awareness of behaviour.

REFLECTION

Andreas Schleicher, Director of the Organisation for Economic Cooperation and Development (2020), suggests that Britain has made the slowest educational progress of all OECD nations. This is because memorisation remains the dominant learning strategy in a narrow, exam-driven culture, which results in neglecting *the personal, transferable competencies* that are a workplace priority and for coping with life situations. He says that education today is not about teaching people something but helping them develop a compass. Work-readiness entails an understanding of *globalisation,* which comes from *mobility* and *partnerships* that require effective intercultural communication. Underpinning education is the ability to process and produce communication for meaning. Since 60% of students worldwide do not reach required educational standards, it is time to review present policies and practices (Luckin, 2020). It is hoped that this book has increased awareness and resolve, particularly of the communicative process, which cannot happen unless both sides are willing to cooperate. In diverse workplaces communication issues have additional complexity. Each culture has tacit assumptions and inclinations in face-to-face verbal exchanges. Attempting to get your message across effectively may be difficult. When language barriers are an issue, communication remains challenging. Therefore, common tips are offered for effective cross-cultural communication that participants in the University of Leicester's Communication Opportunity Groups Strategy (COGS) courses composed for their students in schools and colleges from 2000–2007.

1. Preserve Customs
Cultures have etiquettes about how they communicate. Investigate the target culture before meeting. Many cultures expect formality on a first meeting. This is 'Mr' and 'Mrs' in England, reversing family and given names in China and the use of 'San' in Japan for both men and women. In Britain we shake right hands as a greeting, but in Eastern cultures they bring hands together, as we might in prayer, and incline the head.

2. Speak Slowly
Even when there is a common language, it is unwise to speak at your normal speed. Slowing the pace, with pauses at the end of sentences, will help listeners to translate and process words. Speak clearly and stress important words to enhance meaning. Modulating the pace too much will seem patronising, but it is important to check that information is delivered in a way that others can interpret.

3. Avoid Idioms and Slang
Never assume that any listener knows the same idioms and slang terms that you are in the habit of using. They may understand the individual words but not in the context used. Always check if using a word that may be unfamiliar to others.

4. Use Simple Words and Sentences
Two-syllable words are easier to comprehend than multi-syllabic ones. Short sentences with pauses between gives time for information processing. Say: '*Please everyone, get out your books quickly*', rather than '*Now, everyone, pay particular attention in an efficient manner, with a minimum of noise and distraction, as you need to get out your*

books for writing up your notes on what I have just been presenting to you over the past 30 minutes'.

5. Maintain Turn-Taking
Ensure that one person does not dominate, as taking turns helps a conversation to flow and keeps everyone listening. Make your point and then indicate to the other person that it is their turn to respond. Brief exchanges keep focus on the discussion and are easier to follow.

6. Listen Actively
Restating or summarising the contribution of someone ensures understanding is correct. Ask questions to make sure that you are interpreting the message as it is meant. This promotes active listening and builds rapport, ensuring information is not missed or misunderstood.

7. Promote Open, Not Closed, Questions
Phrasing open questions avoids '*yes*' or '*no*' answers. In countries like Japan it is considered rude to say '*no*', so do not say '*Would you like tea?*' but '*What would you like to drink – I have most most things in stock?*'. This avoids a wrong answer and encourages thinking and talk.

8. Provide a Written Note
If giving information about numbers, such as a time to meet, it is important to write this down on a notelet. It is surprising how often numerical information becomes distorted!

9. Take Care with Jokes
Many cultures believe in behaving formally and professionally – following protocol all the time. If using humour make sure it does not offend and is understood. British dry wit and sarcasm have a negative effect abroad and are not appreciated and may cause offence and hostility.

10. Smile and Support
Conversation only works if everyone feels comfortable. In any situation always treat people with respect and never criticise them negatively, especially if others are present. Discourse with a non-native English speaker can be stressful, so encouragement and support is vital. Smile broadly on meeting to indicate pleasure regarding the prospect of engaging in talk and keep this up to attain a positive outcome!

REFERENCES

Azghari, (2015) *The Bridge Method of Values and Interests*. In Looper, H., *Effective Intercultural Communication*. Copenhagen: Skole & Foraeldre

Holliday, A., Kullman, J. & Hyde, M. (2016) *Intercultural Communication*. London: Routledge

Matteucci, R. (2017) *Communicating the Multi-cultural Classroom: A Challenge for Twenty-first Century Education: Teachers, Students, Families and Administrators*. In Sage, R. (Ed.) *Paradoxes in Education*. Rotterdam: SENSE International

Pinto, D. (2000) *A Three-Step Method for Detecting Difference*. Leuven: Apeidoorn Garant

Sage, R. (2000) *Class Talk: Successful Learning through Effective Communication*. London: Routledge

BIOGRAPHICAL NOTE

Professor Dr Rosemary Sage is a qualified speech and language therapist, psychologist and teacher (English & Maths); Professor of Education (Neuroscience) and former Dean of the College of Teachers. She has been Director of Speech and Language Services, a teacher in primary and secondary schools, a Senior Language Advisor and an academic in five universities, with much national and international experience as a visiting professor and project leader. She is a member of the Judiciary Executive and Magistrates in the Community project. Awards include a Millennium Fellowship for the Communication Opportunity Group Strategy; Fellowship of the College of Teachers; the Human Communication Network International Medal; the Kenneth Allsop Memorial Prize; and College of Teacher Award for the Practitioner Doctorate Model. She is also a non-executive director of the Learning for Life Educational Trust.

Professor Riccarda Matteucci is an experienced teacher at senior school and university levels, having taught in Italy, Africa, America and the United Kingdom. She has held a Research Fellowship in Linguistics at the University of Cambridge. Riccarda also has expertise in across-language teaching, verbal and non-verbal communication and the language-learning problems of those with special needs. She has particular interest in the multi-cultural aspects of students in teaching and the issues that underpin this factor. Riccarda was a Practitioner Doctorate examiner in the area of communication in 2015, at the College of Teachers, then based at the Institute of Education, University College, London. She is presently devising a teacher training programme for African nations.

Professor Barnaby Lenon was brought up on a council estate in south London. He taught at Eton for 12 years, was Deputy Head of Highgate School, Headmaster of Trinity School Croydon and Headmaster of Harrow (12 years). He has been a governor of twenty-two schools and is currently a trustee of the nine independent and state schools in the King Edward's Birmingham Foundation. He is chairman of governors and joint founder of the London Academy of Excellence, a state school which opened in 2012 in Newham, east London. He is Professor of Education at the University of Buckingham, chairman of the Independent Schools' Council and a trustee of the Yellow Submarine charity. He is a member of the Ofqual standards advisory group. He has recently published two books, *Much Promise: Successful Schools in England and Other People's Children* and *What Happens to the Academically Least Successful 50%?* In 2019, he was awarded a CBE for services to education.

Sonny Townsend (artist responsible for the drawings at the start of the book and Chapter 4, p. 120), age 10 years, is in the top class of Malorees State School, in Queen's Park, London. He is interested and talented at music, sports and art. He plays the guitar (Grade 5) and football for Queen's Park Reef Sharks, winning the Players' Player of the Year Award and the Managers' Player of the Year Award for 2018–19.